José Deomar de Souza Barros
Alexson Vieira Pordeus

Social and Environmental Sustainability

José Deomar de Souza Barros
Alexson Vieira Pordeus

Social and Environmental Sustainability

Focus on agricultural practices adopted in communities settled by the São Francisco River Transposition Project

ScienciaScripts

Imprint
Any brand names and product names mentioned in this book are subject to trademark, brand or patent protection and are trademarks or registered trademarks of their respective holders. The use of brand names, product names, common names, trade names, product descriptions etc. even without a particular marking in this work is in no way to be construed to mean that such names may be regarded as unrestricted in respect of trademark and brand protection legislation and could thus be used by anyone.

Cover image: www.ingimage.com

This book is a translation from the original published under ISBN 978-3-330-99616-8.

Publisher:
Sciencia Scripts
is a trademark of
Dodo Books Indian Ocean Ltd. and OmniScriptum S.R.L publishing group

120 High Road, East Finchley, London, N2 9ED, United Kingdom
Str. Armeneasca 28/1, office 1, Chisinau MD-2012, Republic of Moldova, Europe
Managing Directors: Ieva Konstantinova, Victoria Ursu
info@omniscriptum.com

Printed at: see last page
ISBN: 978-620-8-59137-3

ABOUT THE AUTHORS

JOSÉ DEOMAR DE SOUZA BARROS

Degree in Science with a major in Biology and Chemistry; Federal University of Campina Grande - UFCG. Specialist in Agroecology; Federal University of Paraíba - UFPB. Specialist in Chemistry Teaching; Universidade Regional do Cariri - URCA. Master's and PhD in Natural Resources; Federal University of Campina Grande - UFCG. Adjunct Professor at the Federal University of Campina Grande - UFCG.

ALEXSON VIEIRA PORDEUS

Undergraduate student in Biological Sciences at the Federal University of Campina Grande - UFCG. UFCG/CNPq scientific initiation student. Member of the Environmental Research Group for the Sustainable Development of the Semi-Arid - GPA/CFP/UFCG.

SUMMARY

CHAPTER 1

INTRODUCTION

The constant quest to increase productivity and maximise profits has been the keynote of modern conventional agriculture, i.e. based on the precepts of the green revolution, which encouraged the use of chemical agents on plants and the soil, considerably reducing the practice of organic farming, which was widely used before the 1950s, based in general terms on the idea that plants and the soil need to be fed in order to produce more.

If, on the one hand, there has been an increase in the production and industrial development of agricultural implements, on the other hand, there has been an increase in unemployment, rural exodus and concentration of income. The consequences of this degradation have been felt over time (BARROS E SILVA, 2010).

The negative effects of industrial processes on agriculture have led to the emergence of various farming systems that don't use agrochemicals. In this sense, sustainable agriculture has emerged as an alternative and response to industrial urban agriculture. Given that agroecological farming provides for the viability of family farming, associated with aspects of social welfare, food security and the development of local markets (JESUS, 2005).

The adoption of sustainable agricultural practices in semi-arid environments is relevant, given that inadequate management and exceeding the carrying capacity of semi-arid environments have contributed to intensifying the degradation process in locations that are more vulnerable or have more intensified exploitation of natural resources. In this sense, the degradation process is the result of a direct relationship between climatic factors and management mechanisms in the Caatinga Biome (SÁ E ANGELOTTI, 2009).

Thus, in the semi-arid northeast, it is necessary to adopt sustainable agricultural practices in order to increase the adaptability of society and the regional production system to better coexistence with irrigated agriculture provided by the integration of river basins through the São Francisco River Integration Project with the Hydrographic Basins of the Northern Northeast.

CHAPTER 2

OBJECTIVES

2.1. General Objective

To evaluate the agricultural practices adopted in rural communities settled by the São Francisco River transposition project, as well as their socio-environmental, economic and technological aspects.

2.2. Specific objectives

- Observing the agricultural activities carried out by the settled families;
- Checking social, economic, technological and environmental conditions;
- To analyse the obstacles related to the difficulties of implementing sustainable agricultural practices;
- Determine the socio-economic deterioration of the settled communities;
- Assess environmental deterioration.

CHAPTER 3

THEORETICAL BACKGROUND

3.1. Agroecology

The current model of agriculture based on the logic of the Green Revolution proposes development focused on productivity, without the slightest commitment to the rational use of non-renewable natural resources. The negative effects of this model have sparked the construction of a new science: Agroecology, whose basic principle is to support the transformation of conventional agroecosystems into sustainable agroecosystems.

> Faced with an agriculture that increased production and productivity, but which denied natural laws and benefited only certain products and producers, strengthening monoculture, resistance to this process began in the 1920s and especially in the decade after the Second World War. Movements in Europe, the United States and Japan began to revive practices of an alternative form of production that respected the natural principles of agroecology and its various modalities or schools (ZAMBERLAM E FONCHETI, 2012, p. 6364).

In search of alternatives to mitigate the socio-environmental impacts caused by the implementation of technologies in agricultural systems, societies have begun to use versions of sustainable agriculture, such as organic, ecological, biological, biodynamic and others. However, for the most part, these alternatives have not succeeded. Agroecology emerged as a science with the aim of contributing to the construction of ecologically-based agriculture (CAPORAL, 2009).

> [...] Agroecology is emerging as a discipline that provides basic ecological principles on how to study, design and manage agroecosystems that are productive and at the same time conserve natural resources, as well as being culturally adapted and socially and economically viable (ALTIERI, 2012, p. 105).

It was only in 1970 that the term Agroecology began to be used in the scientific environment to characterise an agriculture that opposed the Green Revolution model (ROSA E FREIRE, 2010/2011).

Caporal (2009) points to a problem surrounding the understanding of what agroecology is. For him, even though this term is reminiscent of alternative farming styles, it cannot be confused with a model of such farming. This can be understood in the words of Altieri (2012, p. 16):

> The central idea of agroecology is to go beyond alternative agricultural practices and develop agroecosystems with minimal dependence on agrochemicals and external energy. Agroecology is both a science and a set of practices.

This makes for a complex definition of agroecology, since it is not a type of ecological agriculture, but a science that can collaborate in the study, planning and development of different types of agriculture that are less aggressive to the environment. This complexity is also pointed out by Gliessman (2006, p. 56 apud GUEDES E MARTINS, 2011, p. 73):

> On the one hand, agroecology is the study of economic processes and agroecosystems, and on the other, it is an agent for the complex social and ecological changes that need to take place in the future in order to move agriculture onto a truly sustainable footing.

For Feiden (2005), this science under construction has transdisciplinary characteristics. This is due to the fact that it brings together knowledge from other sciences and disciplines such as Agronomy, Biology, Economics, Sociology, as well as traditional knowledge that is linked to social cultures.

> [...] Agroecology, as a disciplinary matrix, has been laying the foundations for a new scientific paradigm, which, unlike the conventional paradigm of science, seeks to be integrative, breaking with the isolationism of sciences and disciplines generated by the Cartesian paradigm (CAPORAL, 2009, p. 20).

In this sense "[...] agroecology, seen as a simple discipline that studied agroecosystems, has come to be understood in a broader way with contributions from different areas of knowledge" (ZAMBERLAM E FRONCHETI, 2012, p. 75).

The agroecosystem is the object of study of agroecology. It is conceptualised as the modification of a natural ecosystem in order to produce products that guarantee human survival (FEIDEN, 2005). According to the same author,

> [...] the more an agroecosystem resembles the ecosystem of the biogeographical region in which it is located, in terms of its structure and function, the more likely it is that this agroecosystem will be sustainable (FEIDEN, 2005, p. 65-66).

"The main objective of the agroecological approach is to integrate the different components of the agroecosystem in such a way as to increase its overall biological efficiency, productive capacity and self-sufficiency" (ALTIERI, 2012, p. 107).

> [...] agroecology, based on a systemic approach, adopts the agroecosystem as its fundamental unit of analysis, having as its

> The ultimate aim is to provide the scientific bases (principles, concepts and methodologies) needed to implement more sustainable agriculture (CAPORAL, 2009, p. 23-24).

Modern agro-ecosystems, on the other hand, are unstable because they practise monoculture, affecting biological diversity and causing pest outbreaks (ALTIERI, 2012).

For Guedes and Martins (2011), agroecology is a possibility for rural areas, and its practices contribute to families staying in the countryside. It opposes the conventional agricultural model, ignoring the exclusion of peasants and favouring family farming (SILVA, 2010). This contribution is important in society, as it reduces the rural exodus that causes people to congregate in poor living conditions on the outskirts of cities. Silva (2010, p. 4) also points to agroecology as a way of combating the negative impacts caused by the conventional model:

> [...] Agroecology is a major factor in combating food insecurity, mainly by countering inequalities in the distribution of wealth and the conventional agricultural process, which are associated with numerous negative impacts, such as land concentration, violence in the countryside, rural exodus, urban unemployment and unprecedented degradation of natural resources.

The scientific foundations of agroecology promote more sustainable farming styles, recognising the need to produce quality food and contributing to the search for sustainable food and nutrition security (CAPORAL, 2009).

Agroecology not only provides sustainable strategies for agriculture, but it also has a part to play in tackling social problems.

Agroecology seeks a new agricultural approach, conserving local traditional knowledge and using modern ecological methods (ROSA E FREIRE, 2010/2011). "Agroecological systems are deeply rooted in the ecological rationality of traditional agriculture" (ALTIERI, 2012). Farmers therefore have a certain influence on agroecological development. On a day-to-day basis, they observe phenomena in their production systems, and the knowledge they acquire provides information that technicians and researchers would take years of research to obtain (FEIDEN, 2005, p. 68).

> Everyday knowledge is the product of both personal accumulation and the accumulation of successive generations, and its circulation depends directly on memory and wisdom (GOMES, 2005, p. 90 apud ROSA E FREIRE, 2010/2011, p.174).

This recognition of traditional values in the construction of agroecological knowledge can be seen as a regression to the past.

> Building a model of agriculture that respects ecological principles is not a return to the past, as its detractors claim. Although agroecology studies and values traditional agroecosystems, it does so from a critical point of view, in order to understand the logic and interactions that maintain them. It then applies this logic to design new systems that optimise ecological processes and interactions, with the aim of improving the production of useful goods for society (FEIDEN, 2005, p. 66).

That's why it's important to consider that although Agroecology is a science, it doesn't establish theories, but rather looks to lived experiences as a way of studying and designing sustainable development.

> [...] Agroecology does not, for example, offer a theory on rural development, participatory methodologies or methods for building and validating technical knowledge. However, this science looks mainly to the knowledge and experience that has already been accumulated, or through

> Participatory Learning and Action, for example, a method of study and intervention that, in addition to maintaining coherence with its espistemological bases, contributes to promoting the social transformations necessary to generate more sustainable patterns of production and consumption" (CAPORAL, 2009, p. 27).

"Agroecological research, together with agroecological education and rural extension, must articulate the various social forces in the public and private sectors in order to consolidate the urgent need to increase the place of agroecology in the construction of sustainable rural development" (MOREIRA E CARMO, 2007, p. 513). In this sense, there is a need to expand agroecological knowledge in order to promote agriculture that respects the limitations of the environment and the preservation of species as well as human health. "Today, almost all societies are sick because they produce a poor quality of life for everyone, be it human beings or other beings in nature" (ZAMBERLAM AND FRONCHETI, 2012, p.59).

Agroecology is undoubtedly the gateway to a fairer society, one that takes from nature only what it can replenish and uses non-renewable resources consciously. It is also capable of promoting development that values local knowledge and preserves present and future generations. This can be achieved through the implementation of biologically balanced agroecosystems.

3.2. Organic Farming

Recognition of the socio-environmental problems caused mostly by conventional farming practices has increasingly encouraged the spread of agroecological farming models. These include organic farming, the principle of which is to reproduce agricultural systems with minimal use of external inputs.

> Organic agriculture differs from conventional agriculture in that it is socially just, ecologically correct and economically viable. It seeks to promote human health and environmental balance, preserve biodiversity, cycles and the biological activities of the soil. Emphasising the use of management practices that exclude the use of agrochemicals and other materials that perform functions in the soil that are alien to those performed by the ecosystem. Seeking to utilise local resources, thus achieving maximum nutrient recycling (VÁSQUEZ et al., 2008, p. 7).

Organic farming is characterised as a system opposed to the conventional production model, based on agroecology. Its practice involves the management of natural resources, the conservation of agro-ecosystems as well as the production, commercialisation and processing of organic products. In addition, the organic system takes responsibility for health, ethics, citizenship and human autonomy, thus contributing to the preservation of biodiversity (VÁSQUEZ et al., 2008).

For Souza (2015, p. 7), organic farming is conceptualised as:

> [...] a form of agriculture that seeks to act in balance with nature, producing healthy and ecologically sustainable food and products. It is an agroecologically-based agricultural production system that strives to manage the farm as a complex and interactive agricultural organism, with the aim of maximising the flow of nutrients and reducing operating costs.

The adoption of organic farming is initially based on an associative vision between the nature of a place and all its dependencies, relationships and interconnections, as opposed to the ready-made technologies offered by conventional agriculture. The reason for this lies in the logic that each ecosystem has its own particular characteristics, making it unfeasible to use technological packages using only agroecological concepts and principles (PENTEADO, 2003).

The variety of crops, the independence of systems, the conception of the soil as a living organism and respect for nature are the agroecological principles that underpin organic farming (SANTOS, 2012).

Organic production units are considered to be one of the systems that contribute to sustainable development. The more diversified and integrated the agro-ecosystem, the closer it will be to sustainability. Thus, increasing or maintaining biodiversity is the first principle for applying agroecology to the production unit. As such, the goal cannot stop at replacing inputs, but at redesigning agroecosystems (CAPORAL, 2009).

> Organic farms are based on the notion that biodiversity is an integral part of the design of the agro-ecosystem and that, at some point, part of the area will be planted with legumes that will serve as green manures and will be incorporated into the soil or will serve as fodder for livestock, whose manure

will be returned to the soil (ALTIERI, 2012, p. 63).

Replacing conventional inputs with organic ones is considered a relevant stage in the process of transitioning to agroecological production. However, this substitution cannot be considered the final stage, given the objective of the sustainability of the agroecosystem in its economic, social, ecological and agronomic dimensions (ASSIS, 2006).

The organic production system excludes the use of synthetic inputs such as fertilisers, pesticides, growth regulators and food additives. This is achieved by adopting crop rotation practices, recycling waste, green manures, mineral rocks, management and biological control. The aim is to maintain the fertility and micro-life of the soil (PENTEADO, 2003). "In organic farming, the soil is fertilised, i.e. the micro-life is fed, not the plant. The plant is fed by the micro-life" (ZAMBERLAM AND FRONCHETI, 2012, 87).

In this sense, organic fertilisation is important for maintaining soil fertility and increasing yields. Organic fertiliser and green manure are important practices for agricultural management .

> [...] green manure is a natural process that aims to protect the soil's surface and improve its chemical characteristics. And by adding natural products, it is possible to have good quality agricultural fertiliser with reduced risk for the farmer. Animal manure is another form of fertiliser that has the benefits of cleaning the animal's sleeping place, restoring nutrients to the soil to be cultivated and reducing the risk of explosions due to the accumulation of gases (SANTOS et al, 2014, p. 3).

According to the International Federation of Movements of
Organic Agriculture - INFOAM, the organic system is practised in over a hundred countries. In the European Union, approximately 80,000 properties produce organically; in the USA, around 1% of the food market is organic; in South America, Argentina stands out as the largest producer of organic food and in Brazil, around 100,000 hectares, spread over 5,000 production units, adopt the organic agricultural

system (VÁSQUES et al., 2008).

Currently, organic farming in Brazil is concentrated on supplying products for direct consumption, particularly dairy products, tinned foods and fresh vegetables. These products are concentrated in the south and southeast, and are sold at fairs and natural products shops, with a growing increase in consumption. The south-east has also been the biggest consumer of organic products in the country, but it is important to emphasise that the Brazilian market for organic products is expanding throughout the country. This expansion is due to factors that motivate consumers to buy organic products. These include the exclusion of the use of agrochemicals, the preservation of the environment and personal and family health, the flavour of the product and its biological value (SANTOS, 2012).

It is estimated that 90 per cent of Brazilian organic farmers are classified as small producers, linked to associations and social groups. The other 10 per cent are classified as large producers linked to companies. In terms of production, 70 per cent of organic products are produced by family farmers (BORGUINI E TORRES, 2006, p. 66 apud SANTOS, 2012).

According to Penteado (2003), organic farming is a system that aims to produce food with its original characteristics and flavour, thus meeting consumer expectations. In this sense, Souza (2015) emphasises that the quest to improve quality of life is a fundamental factor in expanding the consumption of organic products.

> Today, consumers around the world are increasingly demanding 'clean' products that are free from chemical substances and are not genetically modified. They are therefore seeking a better quality of life by prioritising healthy eating. This concept has influenced the development of organic farming (SANTOS, 2012).

On the other hand, Vásquez et al. (2008) point out that there are several inhibitors to increased consumption of organic products, including high prices, little variety and a lack of information on the part of consumers. Consumption should therefore be stimulated through strategies aimed at increasing knowledge and consequently stimulating purchases.

> In order to increase the value of their product, organic farmers must seek to formalise a certification system to obtain a label for their organic product. Farmers who have the production conditions throughout the life cycle established by this certification system will have access to niche markets with higher rates of return for their product, associated with a guaranteed sales regime and the construction of a quality image with their customers (VÁSQUES et al., 2008, p. 9).

With this, labelled organic products awaken consumers to recognise their quality and their ability to contribute to food safety. This guarantees the expansion of the organic food market and the maintenance of their production systems.

3.3. Sustainable Rural Development

The critical analysis of agricultural production systems has been disseminating the proposal for development that allows for harmonious interaction between farmers and their rural environment, the agro-ecosystem. This development is based on recognising the limitations of natural resources, given the high levels of degradation caused by unsustainable agricultural practices.

The conceptual basis for sustainable development emerged at the Stockholm Conference in 1972, and was initially named the "eco-development approach" (DIAS, 2009). It was only in 1987, with the publication of the Brundtland Report - Our Common Future, that the term sustainable development was established. This report officially recognised the limitation that the environment imposes on growth, proposing a "green" side to current economic development (PATRÍCIO E GOMES, 2012).

In Brazil, concerns about the consequences of industrial agriculture, based on the Green Revolution, began to emerge in the mid-1970s and grew stronger in the early 1990s when the first results of initiatives to mitigate the socio-environmental problems caused by industrial agriculture were seen. During this period, ecological sensitivity arose from the realisation of the deterioration of the planet and the imbalances in existing ecosystems (ASSAD E ALMEIDA, 2004).

> From the 1970s onwards, the results of the application of conventional development strategies were already proving insufficient to deal with the growing conditions of inequality and social exclusion. Despite GDP growth, analyses of these results began to indicate that these strategies were causing serious damage to the environment. The contaminating effects of pesticides, waste, rubbish and gas pollution, as well as various other problems deriving from the lifestyle of highly industrialised societies, for example, raised awareness of the inability to control the externalities inherent in the hegemonic model and, therefore, the need for 'another development' (CAPORAL E COSTABEBER, 2000, p. 19).

In this sense, "development, in its broadest formulation, would mean the realisation of a society's socio-cultural and economic potential in perfect harmony with its environmental surroundings" (CAPORAL E COSTABEBER, 2000, p. 18). This is in contrast to the concept of development imposed by liberal thinking, which associates it with an idea of economic growth defined by standards of living and consumption. In this way, development is represented through the race of different societies towards a socio-economic model considered 'developed', moving from an unworthy condition, underdevelopment, to a model of capitalist society (CAPORAL E COSTABEBER, 2000).

> As an alternative, a development process has been sought that is based on qualitatively distinct economic growth and that makes it possible to maintain or increase, over time, the set of economic, ecological and socio-cultural goods, without which economic development is not sustainable, i.e. it is necessary to combine social justice and the conservation of natural resources independently of economic growth (ASSIS, 2006, p. 80).

According to Zamberlam and Froncheti (2007, p. 85):

> A development process is sustainable when it achieves the fulfilment of needs without compromising natural capital and without harming the right of future generations to have their needs met and to inherit a healthy planet with its ecosystems preserved.

For the concept of sustainable development to be implemented, it is necessary to aim for harmony and rationality not only between man and nature, but above all between human beings. They must be agents in the development process, which must be achieved with respect for ethnic characteristics, aiming to improve the quality of life of populations, with an emphasis on the poorest (ASSIS, 2006).

The sustainability of an agro-ecosystem can be seen through two essential components: environmental and social. Environmental sustainability is related to the effects that agroecosystems have on renewable and non-renewable natural resources, such as their collaboration in the processes of contamination, global warming, deforestation and other factors that harm the environment on both a global and local scale; social sustainability is related to the internal capacity of agroecosystems to overcome the external pressures to which they are exposed. This capacity guarantees the fulfilment or not of the socially desired objectives for satisfying human needs (FERNÁNDEZ E GARCIA, 2001).

According to Patrício and Gomes (2012), the local level is the fundamental space for promoting Sustainable Rural Development, since it is there that the community develops actions to resolve problems of production, consumption, occupation and local utilisation of its potential. From this perspective, Araujo and Arruda (2011, p. 243) emphasise that "sustainable development transforms the locality that produces with ecological awareness and with a focus on human interaction and its political and social construction".

> [...] sustainable rural development must be based on participatory planning whose guidelines are guided by respect for: the productive potential of ecological systems; the use and occupation of space; the production of goods aimed at basic social needs; cultural values; and the production of social wealth with a focus on participatory community management for self-determined endogenous development (PATRÍCIO E GOMES, 1012, p. 105).

The conventional model of rural development causes economic, social and ecological problems, contributing to the spread of a poorly competitive agriculture that is unable to guarantee food security, with homogeneous management systems that run counter to ecological principles and the aim of producing renewable resources (food) through the use of non-renewable resources (fossil fuels), contributing to the process of degradation (FERNÁNDEZ E GARCIA, 2001).

In opposition to this scenario, the sustainable agriculture movement proposes technologies that imply a certain break with the conventional model, valuing practices adapted to farmers (ASSAD E ALMEIDA, 2004).

> [...] the term sustainable agriculture refers to the pursuit of long-term, durable yields through the use of ecologically sound management technologies, i.e. ecologically based farming styles that meet the requirements of solidarity between current generations and from these to future ones (ZAMBERLAM AND FRONCHETI, 2012, p. 75).

The 1970s saw the emergence of agroecology, a science aimed at providing a theoretical basis for the different movements of non-conventional agriculture. Agroecology seeks to understand how agroecosystems work, with the aim of conserving and expanding their biodiversity (ASSIS, 2006). Caporal and Costabeber (2000, p. 26) follow the same line of thought when they emphasise that "[...] Agroecology provides the scientific basis to support the transition process to Sustainable Agriculture styles in their various manifestations and/or denominations: Ecological, Organic, Biodynamic, Agroecological, Regenerative, Low External Input, Biological, among others."

> [...] agroecology is a way of guiding development in agriculture (at the level of isolated niches) in a harmonious way, as it is based on ecological, social, cultural, spatial and economic principles of sustainability, which make it possible to interrelate all these principles efficiently (PATRÍCIO E GOMES, 2012, p. 104).

3.4. Family farming

Family farming can promote practices related to Sustainable Rural Development, given that family production stands out as the main economic activity in Brazil. Such production contributes to the development of sustainable agriculture due to its diversified practices, which integrate plant and animal activities with smaller scales of labour (SOARES et al., 2009).

According to Soares et al. (2009, p. 57), "family farming can be defined as the set of agricultural production units operated on a family economy basis, comprising those activities carried out on small and medium-sized properties, with the family's own labour force". Zamberlam and Froncheti (2012, p. 54) point out that "for public bodies, the concept of family farming covers that area of the rural establishment or enterprise that does not exceed four fiscal modules, where the labour force in economic activities is predominantly family-owned and the enterprise is run by the family".

The implementation of sustainable practices in agricultural systems requires complex changes to the development model in force in today's society, with the development of alternatives based on local and regional realities (ASSIS, 2006). As a result, the countryside will be handled in a diversified way, guaranteeing biodiversity and food security, thus collaborating towards economic development that unifies production and the rational use of natural resources - Sustainable Rural Development.

3.5. The Brazilian semi-arid region

The characterisation of the Brazilian semi-arid region allows for a complex understanding of its natural, biological and social aspects. From then on, the semi-arid region began to be seen as a diverse environment, disregarding existing misconceptions and a new concept emerged: coexistence with the semi-arid region.

The Brazilian semi-arid region covers an area of 969,589 kilometres^2 and includes 1,133 municipalities with a total average population of 28 million. In terms of location, the Semi-Arid covers the central part of the northeast region with the states of Piauí, Ceará, Rio Grande do Norte, Paraíba, Pernambuco, Alagoas, Sergipe, Bahia and part of the state of Minas Gerais, in the southeast region (BARROS, 2014).

> When people talk about the semi-arid region, one issue immediately comes to mind: water, rain and drought. It is often said that it doesn't rain enough, that there is a lack of water and that this is the biggest problem in the semi-arid region. This is a relative truth, as there are marked differences in annual rainfall from one region to another. [...] Ours is the wettest semi-arid region in the world, but rainfall is concentrated in just a few months and more than 90 per cent of its water is not used due to evaporation and surface run-off (BAPTISTA E CAMPOS, 2013, p. 47).

The water sources in the Brazilian semi-arid region are insufficient to meet the demands of the population. The reasons for this are strong sunshine, high temperatures and irregular rainfall (SILVA et al., 2010).

The scarcity of perennial rivers is due to the geological structure of the soils, which have crystalline characteristics and are therefore shallow, making infiltration difficult and causing surface runoff. In addition to this natural problem, human action causes degradation of the vegetation cover that protects them from erosion (ARAÚJO, 2011). According to Brasileiro (2009, p. 5), "If the native vegetation cover is maintained, the possibility of any degradation is small, and degradation due to anthropogenic causes is even smaller".

According to Silva et al. (2010), it is not the lack of rainfall that is the limiting factor for the precarious availability of water in the semi-arid region, but rather its poor distribution in conjunction with high evapotranspiration and the lack of public policies to guide populations towards appropriate ways of capturing and storing rainwater for use during the dry season.

Against this backdrop, it can be seen that climate is a key element in characterising the semi-arid region. It determines the adaptation of vegetation, as well as the formation of relief and the impermeability of the soil (ARAÚJO, 2011). The phenomenon of droughts is a case to be considered within the climatic problems of this region. Its characteristics include the depletion of soil moisture, the fading of plants and a reduction in the flow of watercourses. In addition to these, the tragedy of drought causes serious social, economic and political problems, with a reduction in agricultural activities, a lack of water for human consumption, causing illnesses and even deaths from drinking contaminated water (BAPTISTA E CAMPOS, 2013).

The predominant biome in the semi-arid region is the Caatinga, characterised by a semi-arid tropical climate and a diversity of environments that make up large areas of land in the northeastern interior (BRASILEIRO, 2009). According to Barros (2014), despite its aridity, the Caatinga is a complex ecosystem, as the living beings in this ecosystem have adaptive potential to drought conditions. Plants, for example, have adapted in such a way that they are able to take advantage of the minimal amount of water in this environment. According to Angelotti et al. (2011, p. 1,104) "adaptation refers to the adjustment of natural or human systems in response to observed or predicted climatic stimuli, with the aim of increasing the resilience of these systems".

The Caatinga has 80% of its area altered due to predatory human exploitation, leaving it with a high potential for vulnerability to climate change (ANGELOTTI et al., 2011). This is because the Caatinga conditions human activity in this region, providing wood for the production of firewood, charcoal, building materials and more. In addition, fruits, medicinal plants and honey are of economic importance to the population. The vegetation is also used to produce fodder to feed extensive livestock farming during the dry season. These practices of exploiting the Caatinga's natural resources without proper management contribute to its degradation, accentuating the rural exodus (BARROS, 2014). Brasileiro (2009) also points out the various factors that are contributing to the Caatinga's degradation process, including inadequate agricultural practices, deforestation and soil infertility, compaction, erosion and salinisation.

> [...] the degree of soil cover in the caatinga has not always been what we have today; its vegetation has suffered a process of degradation due to the use of firewood, constant burning and deforestation for land use in agriculture, making it possible for various species to disappear or thin out - which is a considerable loss, as the biodiversity of this ecosystem is very rich (ARAÚJO, 2011, p. 93).

Thus, we can understand why the semi-arid region is so arid, highlighting some of the factors that contribute to this occurrence, including human ways of exploiting the land, making it desert, deforestation, predatory practices on rivers and land, fires and soil contamination by pesticides. All of these processes are combined with the scarcity of rainfall and the inability of water storage systems (BAPTISTA E CAMPOS, 2013).

According to Brasileiro (2009), the areas vulnerable to the desertification process are considerable and are one of the reasons for the worsening environmental impacts in the semi-arid region. One of the problems in defining the desertification process is associated with the lack of monitoring and evaluation policies for this process, thus causing the occurrence of unsustainable soil and land management mechanisms (SANTOS, 2011).

Studies carried out in the semi-arid region show that anthropogenic interference with the environment, negative processes on the fauna, flora and soils, as well as the climate, constitute significant rates of desertification. For this reason, it is important that the natural aspects of this region are considered and understood, so that there can be better coexistence in this environment (SILVA et al., 2010).

Barros (2014) points out that the process of degradation in the Brazilian semi-arid region is not just a consequence of the natural conditions of the region, but mainly due to the use imposed on it. This is why it is necessary to adopt practices with the potential to contain and reverse the deterioration process, through a broad programme of coexistence with the semi-arid region.

> In the broadest sense, the expression Coexistence with the Semi-Arid cannot be understood only in terms of the possibilities of adapting to the geo-environmental particularities of the Dry Lands, which are interconnected with the global processes of the current climate change situation, such as periodic droughts and the intensification of their frequency, because it is a question of understanding the web of complex relationships between human and natural systems (SANTOS, 2011. p. 165).

In this way, the term Coexistence with the Semi-Arid emerges as a new way of looking at the relationship between man and the natural environment, seeking socio-environmental sustainability.

In addition to environmental problems, the Semi-arid region has worrying economic and social indicators. Economic activities show inequalities in land distribution and income concentration (BARROS, 2014). "The semi-arid region has the highest levels of socio-economic vulnerability, with a large part of the population engaged in agricultural activities [...]" (ANGELOTTI et al., 2011, p. 1098). Mitigating these socio-economic problems is a challenge for the development of the semi-arid region, through economic opportunities that generate jobs and local income (SILVA et al., 2010).

The manifestation of these negative adversities in the Brazilian semi-arid region, especially the ways in which man exploits the natural resources available, are among the main causes of degradation in this region. The climate, vegetation, soil and, above all, water sources are being affected in a disorganised way, causing great damage to the environment and the population of this region. In this sense, the adoption of sustainable practices tends to contribute to a better coexistence between man and nature.

3.6. Agriculture in the Semi-Arid

Inadequate handling of agro-ecosystems in the semi-arid region is among the main anthropogenic activities causing the deterioration of this region. Deforestation, burning and the uncontrolled use of chemical products are among the practices most used by farmers in food production, causing unproductivity and soil erosion, the main component in agricultural development, as well as climate change.

Pereira et al. (2010) point out that food production occurs through two types of agricultural systems: industrial and traditional. Industrial agriculture is practised using large amounts of energy from fossil fuels, water, fertilisers and pesticides; traditional agriculture is subdivided into two groups: traditional subsistence agriculture, practised only for the survival of the producing family using family labour and beasts of burden, and traditional intensive agriculture, which as well as producing for subsistence also generates income by increasing labour, resulting in an increase in production.

The farming system in the semi-arid region is characterised as traditional rainfed, using slash-and-burn soil preparation. In this model, farmers have a certain ability to interpret natural signals in order to plan their farming calendar, determining the start of land preparation, planting and harvesting activities (NASUTI et al., 2013).

Agricultural practices in the northeastern semi-arid region vary greatly. This variation is related to both the crops planted and the technologies used to produce them. Sugar cane stands out as the region's main agricultural product, followed by cotton, soya, maize and others (CASTRO, 2012).

"Agriculture in itself is an activity that generates a lot of impact on the environment, whether on a large or small scale; this will depend on the techniques and practices that are used to cultivate the land" (BRASILEIRO, 2009, p. 4). Thus, the process of degradation in the semi-arid region begins with agricultural practices that remove the soil's original vegetation cover, without properly replenishing nutrients (BRASILEIRO, 2009). Silva and Rios (2013, p. 5) mention this practice as the main cause of soil erosion:

> Inadequate soil management and the destruction of vegetation cover facilitates the process of soil erosion. As a result of erosion, soils need more nutrients, which are not always replenished satisfactorily to meet the nutrient needs of plants. This causes farmers to move from agricultural areas to wooded areas, thus starting the cycle of degradation.

In addition to erosion, contamination and compaction are other factors to be considered among the causes of soil deterioration in the semi-arid region. The model of agriculture based on productivity, with the intensification of food production, causes contamination of the soil by agricultural inputs, also affecting the air and water. The excessive mechanisation of this model affects the structure of the soil, causing it to become compacted and preventing the plant's root system from attaching (PEREIRA et al., 2010).

According to Silva and Rios (2013), the agricultural practices developed in the semi-arid region are knowledge passed down from generation to generation and are often inadequate. The high rate of illiteracy is a barrier to the acquisition of knowledge for the majority of farmers in the northeast, preventing this reality from changing. This results in impoverished soils, with a lack of vegetation cover and signs of desertification. In order to alleviate this situation, it is necessary, according to Silva and Rios (2013, p. 4) "[...] to inform, train and sensitise farmers about these problems, since knowledge of the issue can help to reduce the problem".

Edaphoclimatic variations with the problem of drought, the high cost of transporting goods, the use of outdated technologies and poor production storage mechanisms are other aspects that limit the development of agriculture in the semi-arid region (CASTRO, 2012).

Nasuti et al. (2013), when carrying out research in semi-arid regions, found that 70% of the farmers interviewed classified climatic factors and difficulties in accessing water as an obstacle to agricultural production. And around 60 per cent adopt the rainfed system, being totally dependent on rainwater. As a result, the dissemination of rainwater harvesting reservoirs has become one of the most viable alternatives for farmers who do not have perennial watercourses.

For these reasons, the development of agriculture in the semi-arid region needs to be practised in a sustainable way, adopting correct techniques for managing and conserving soil and water, thus increasing production without exhausting the land (PEREIRA, 2010). Among the alternative experiences being adopted in semi-arid regions, sustainable agriculture stands out the most. However, this alternative requires radical changes to the current family farming system. Agroecological practices are innovative because they bring about a change in behaviour in the relationship between the sertanejo and the vegetation (BRASILEIRO, 2009).

> The agroecology developed in the semi-arid region is consolidated to the extent that family farmers are nourished by a deeper vision of their relationship with the environment. It seeks to combine farmers' knowledge and experience through a closer relationship between their concepts and methods and sustainable development practices. This whole process gives the territory of the caatinga, and more specifically small family farms, a new spatial configuration, as these territories take on a new productive and sustainable function. They acquire a new meaning for farmers by preserving cultural roots, maintaining the natural resources of the caatinga biome and adding value to the work of family farmers, since agroecological products can be sold on a differentiated market (BRASILEIRO, 2009, p. 9).

In this sense, it is important to adopt sustainable agricultural practices for the productive development of the semi-arid region, given the lack of effective water systems, fertile soils and adequate public policies for this region. In addition, the implementation of such practices contributes to the sustainable management of Caatinga resources, mitigating the impacts of degradation and consequent desertification.

3.7. From combat to coexistence: a paradigmatic transition

Water scarcity has always been seen as a limit to development in the semi-arid region, leading to the establishment of an inefficient policy to combat drought. The social problems affecting the semi-arid region have always been associated with natural phenomena, and public policies, rather than mitigating them, are often used to strengthen the drought industry, concentrating public resources in the hands of a minority. Faced with this scenario, a new paradigm has emerged in recent years: coexistence with the semi-arid region. This aims to help improve the relationship between man and his environment, through environmental education and the development of sustainable technologies that make it possible to promote a dignified life for the people of this region.

Combating drought was, and to some extent still is, adopted as the basic policy for the semi-arid region (BAPTISTA E CAMPOS, 2013). In this paradigm, drought appears as a natural obstacle, a phenomenon that must be combated, and is therefore a reductionist logic that maintains misery. The fight against drought follows the same pattern of distancing between man and nature formulated by modern society; its aim is to tame nature, i.e. drought and its effects (SILVA, 2003).

Coexistence with the semi-arid region is beginning to emerge as a new paradigm based on an ecological vision, breaking away from the anthropocentric vision and reconciling man with nature, based on a complex perception of the reality of ecosystems and the valorisation of knowledge, values and practices concerning the environment (SILVA 2003). This new reality that the Brazilian semi-arid region has been building shows that, just as it is possible to live on ice, it is also possible to live in semi-arid and arid regions, through strategies for living with these regions. In this sense, drought ceases to be a problem and becomes a challenge, to be overcome through appropriate public policies (ALVES, 2013). Malvezzi (2007, p. 11-12) also emphasises the idea of coexistence with the semi-arid region:

> A new concept of civilisation is being developed for the region: coexistence with the semi-arid region. The idea is based on a simple principle: why is it that the people of the ice can live well on the ice, the people of the desert can live well in the desert, the people of the islands can live well on the islands and the people of the semi-arid region live badly here? It's because those peoples have developed cultures of coexistence suited to the environment, adapted to it and made life viable. In the Brazilian semi-arid region, this integration of people and nature has not found an adequate solution, so that human beings have remained subject to the normal variations of the regional climate.

This paradigmatic transition based on coexistence and development with quality of life in the Brazilian semi-arid region requires a combination of social, economic, cultural and political actions, linked to values and practices that respect human dignity and the dignity of other living beings. Democratisation and access to quality water, access to land, the promotion of contextualised education aimed at coexistence, incentives for productive activities and access to basic services for the population are among these actions (CONTI and PONTEL, 2013).

The widespread image of the Semi-Arid has always been distorted, creating the idea of an arid region rather than a semi-arid one, as if it didn't rain, as if the soil was always calcined, the forests dry and the lack of rain was constant (MALVEZZI, 2007). However, as Duarte et al. (2015) state, the Semi-Arid is not just about climate, it is also about vegetation, soil, people, music, festivals, religion and other socio-cultural aspects.

> The semi-arid region is not an unproductive space. Native animals, plants and trees grow and live there. People plant and harvest. It rains in the semi-arid region, if not abundantly, but enough to ensure life. It turns out, however, that much of what is produced and what nature makes available in the semi-arid region is not sufficiently utilised, due to the lack of a culture that creates the conditions to store what is produced in times of abundance in order to use it in times of greater need and thus guarantee life and food security (BAPTISTA E CAMPOS, 2013, p. 54).

In this way, rainwater harvesting emerges as the first law of coexistence with the semi-arid region, a practice used since biblical times by the people of Israel, which was almost abandoned due to the large amount of water in Brazilian territory. Recently, this practice has been guided by the National Water Resources Plan, whose policy is to capture rainwater for human and animal consumption and for agriculture. In addition to water, it is also necessary to store goods produced in rainy times for use in times without rain (MALVEZZI, 2007).

Coexistence requires a change in the perception of territorial complexity, rescuing and building a relationship between humans and nature, with the aim of improving the quality of life of populations living in semi-arid regions. This new way of perceiving the semi-arid region based on its particular characteristics, its limits and its potential eliminates the "blame" attributed to the region's natural conditions, thus guaranteeing its development (CONTI & PONTEL, 2013).

According to Santos (2011), adapting and managing are the watchwords for living in the semi-arid region, maintaining a balance between social activities and natural potential, while also giving importance to popular knowledge. Baptista and Campos (2013) also follow this line of thought when they state that living together means living, producing and developing, without promoting the concentration of goods, but rather sharing, justice and equity. In contrast to the model of combating drought.

Duarte et al. (2015), when carrying out studies on the coexistence actions developed by the Ceará State Department of Education, indicated that the inclusion of environmental education contributes to better coexistence in semi-arid regions. "Human interference in the environment is inherent to the very condition of life of individuals. However, the way in which this interference is practised is worrying, hence the importance of environmental education as an instrument for improving man's relationship with his habitat" (DUARTE et al. 2015, p. 20). In this sense, they emphasise that contextualised teaching would be a viable alternative for coexistence with the semi-arid region, by promoting environmental education that is consistent with the reality of each region.

> [...] the construction of knowledge about the semi-arid region makes it possible to develop projects and actions to coexist with regional environmental conditions. A curriculum that dialogues with the knowledge and practical experiences of individuals, enabling them to develop a critical awareness of their role as subjects in building a better semi-arid region to live in (DUARTE et al., 2015, p. 25).

For Silva (2006), coexistence with the semi-arid region is also a political proposal aimed at mobilising civil society and the state to implement public policies that promote sustainable development in the region. In this sense, it is worth noting that part of the initiatives are conveyed to the population, through the mobilisation and organisation of social movements that make it possible to spread the social values of coexistence and push for improvements in economic and social conditions.

> Living with the semi-arid region means reorienting the gaze of public managers in the search for sustainability, starting from the community's own vision of development, which has been gradually overshadowed by the drive for growth and/or economic development at any cost (even human survival), defended by the world's major powers (SANTOS, 2011, p. 166).

In this context of mobilisation for coexistence with the semi-arid region, it is important to highlight the important collaboration of NGOs and research institutions in developing viable technologies for the reality of semi-arid regions. Silva (2003) points out that the construction of this new paradigm has its origins in the initiatives of research centres, such as the Brazilian Agricultural Research Corporation (Embrapa), and NGOs, which have been developing projects in areas of the Brazilian semi-arid region since the 1980s. At the end of the 1990s, the proposal of coexistence was reinforced through the creation of the Semi-Arid Articulation (ASA).

The ASA is a forum for civil organisations with the same goals: to discuss and propose coexistence policies for the semi-arid regions (SOUZA, 2014). Its aim is to develop policies

sustainable coexistence, fighting for the social, economic, political and cultural development of the peoples of this region, as well as denouncing corrupt and unviable practices that may affect them (ALVES, 2013).

The paradigm shift from combating drought to living with it has exposed the misconceptions that hinder development in the semi-arid region, promoting the population's criticality and commitment to promoting technologies that are appropriate to the local reality. Living with the semi-arid region means creating possibilities for development, as opposed to combating drought, which imposes obstacles to people adapting to the region.

3.8. Natural resources: a sustainable approach to living in the semi-arid region

The current development model imposed by modern society is aimed at profit, achieved through mechanisation and inadequate handling of available natural resources. Contrary to this thinking, sustainable development, according to Lira and Oliveira (2012, p. 277) "[...] appears as a new model that reconciles wealth and development with environmental preservation and the fight against poverty". In the semi-arid region, the sustainable use of the Caatinga's resources contributes to its conservation as well as to the construction of coexistence appropriate to this ecosystem.

Sustainable development is based on the need to promote economic development by satisfying the needs of the present generation without compromising future generations (ARAUJO E ARRUDA, 2011). Lacerda and Cândido (2013) follow this same line of reasoning when they conceptualise sustainable development as the search for limits to the system of economic development with the aim of meeting interests of current humanity while preserving and conserving existing resources in order to guarantee the livelihood of future generations.

In the Brazilian semi-arid region, building a sense of coexistence is a fundamental challenge for the new guidelines for sustainable development in this region. Coexistence is not simply about developing new techniques, new activities and productive practices, but a cultural proposal, with the aim of contextualising knowledge and practices appropriate to the semi-arid region, taking into account its diversity, popular knowledge, problems and alternatives that have been built up over the history of its occupation (SILVA, 2006).

According to Silva and Rios (2013), the main cause of natural resource degradation is negative anthropogenic activities, especially inadequate agricultural practices. In this sense, they highlight the agricultural techniques used by farmers in soil management, causing the depletion of its nutrients and organic matter, leaving it prone to erosion, acidity and salinisation. Deforestation and burning also contribute to worsening this situation. According to Brasileiro (2009), the removal of native vegetation cover is one of the main indicators contributing to the processes of degradation and desertification in the semi-arid region. Soil fertility is important for the sustainable development of this region, given the small amount of land suitable for cultivation. In relation to soil erosion, Brasileiro (2009, p. 5) emphasises that:

> [...] erosion is one of the most serious problems on the degradation scale, because it generally causes irreversible impacts on the environment. As far as the Northeast is concerned, more precisely the semi-arid region, erosion processes are a cause for concern, as the soil is increasingly vulnerable due to intensifying anthropogenic action and the very fragility of the pedological material: shallow, gravelly and often sandy soils.

An effective practice in mitigating erosion processes and the loss of nutrients that guarantee soil fertility is the use of methods with minimal disturbance, crop rotation systems, preservation of native vegetation and reforestation, which contribute to increased productivity and the accumulation of plant residues, resulting in an increase in carbon stocks in the soil (ANGELOTTI et al., 2011).

Over the last five decades, the process of land degradation, due to disorderly deforestation and rudimentary farming practices, has increased uncontrollably, leading to the disappearance of many native species of fauna and flora. In addition, another consequence of irrational deforestation is the transformation of regions in Brazil and around the world into veritable deserts, especially in the semi-arid region, where anthropogenic activities have caused a marked degradation of natural resources, giving rise to desertification centres (DUARTE E BARBOSA, 2009). According to Araújo et al. (2011, p. 257) "desertification can be considered as the final expression of the process of land degradation and depredation of natural resources and, in this sense, it is human actions that are primarily responsible for the implementation of this process".

According to Barros (2014), the adoption of sustainable techniques in the Caatinga is capable of diversifying production, thus ensuring the sustainability and containment of the devastation of the semi-arid vegetation. Silva (2006) also states that sustainable management of native vegetation, combined with changes in the energy matrix and agricultural practices, reduces deforestation in the Caatinga biome, especially in regions prone to desertification processes.

> Coexistence with the environment is a fundamental imperative for the sustainable management and use of natural resources in an ecosystem, without making their reproduction unfeasible. It implies a new orientation for human activities, seeking to reconcile or correct the natural limits to human intervention (SILVA, 2006, p. 227).

Climate information is of fundamental importance for the rational management of the Semi-Arid's natural resources (SILVA, 2010). The water scarcity that affects this region is not related to a lack of rain, since the Brazilian semi-arid region is considered the wettest on the planet, with average rainfall of between 250 and 800 mm/year. The problem is that rainfall is lower than evaporation levels, which is 3,000 mm/year (MALVEZZI, 2007).

The implementation and dissemination of social technologies for storing water resources has ensured that the population of the semi-arid region can make use of the water stored during the rainy season, taking into account the high evaporative demand. Articulação no Semiárido Brasileiro (ASA) has been implementing and testing around forty social technologies, some of which have already become programmes, such as the "One Million Cisterns Programme (P1MC)" and the "One Land and Two Waters Programme (P1+2)". As well as investing in the dissemination of social technologies, it is also necessary to promote the need for shared and responsible management of the water resources available in the semi-arid region (ARAÚJO et al., 2011). As Silva (2006, p. 230-231) points out:

> In addition to appropriate water collection and storage technologies, living in the semi-arid region requires community management of water sources. Shared management of this natural resource is a necessary social and political initiative to guarantee the sustainable use of water, enabling human supply and appropriate production without degrading surface water sources and underground aquifers.

According to Malvezzi (2007), the concept of social technologies arose from critical thinking about technologies in general, given that the current reality demands that they be sustainable in order to ensure the good of all. Among the social technologies that are being disseminated in the Brazilian semi-arid region, the ones that stand out the most are spouted cisterns and wedge cisterns. The latter is designed to collect rainwater for production by means of a cemented pavement in the ground, while the latter is designed to provide quality water for human consumption by collecting water through the roof of the house. Both are hermetically sealed, allowing no light in and no living organisms to multiply.

> Social technologies have helped to build another scenario in the semi-arid region, firstly because they make it possible to recreate ways and

> possibilities of living with the difficulties of drought, without losing sight of the accumulation of knowledge that producers and users of these technologies have learnt, leading them to value their entities, but also innovating their practices, thus making it possible to build life projects based on sustainability (SOUZA, 2014, n/p).

However, if there is no water management and treatment, all water alternatives can become inefficient. It is therefore necessary to implement systematic awareness-raising and training processes for the correct management of water resources. One example is the water management courses held with the families benefiting from the cisterns. They help to mitigate waste and contamination of stored water, and participants are encouraged to practise sustainable management of local water resources by cleaning up existing dams and reservoirs and preserving or replenishing riparian vegetation (SILVA, 2006).

> [...] water resource management with a view to sustainability seeks to implement a set of actions aimed at regulating the use, control and protection of water resources, in accordance with the relevant legislation and standards. It is also necessary to integrate projects and activities with the aim of promoting the recovery and preservation of the quality and quantity of river basin resources, as well as the recovery of springs, water sources and watercourses in urban areas (LACERDA E CÂNDIDO, 2013, p.16).

The sustainable management of natural resources, especially water as it is an indispensable resource for human survival, is of fundamental importance for the development of the semi-arid region, as it not only conserves the Caatinga biome but also helps to disseminate practices for living together in this region.

CHAPTER 4

METHODOLOGICAL PROCEDURES

The activities took place between August 2015 and July 2016. They were guided by qualitative and quantitative research, and a case study was used as the methodological strategy.

4.1. Location and Description of the Study Area

The transposition of the São Francisco River will take place in the area of the Polígono das Secas, more specifically in the northern north-east, and is a water infrastructure project. Two independent systems called the North and East Shafts will draw water from the São Francisco River. The catchment at Cabrobó will start the so-called Northern Axis, which, due to its length, has been divided into five stretches: Stretches I, II, III, IV and VI.

The research was carried out with families settled in the productive villages assisted by the São Francisco River Transposition Project in the Piranhas River Sub-basin, Section III, in the city of São José de Piranhas - PB.

4.2. Research Classification

The methodology adopted by Barros and Silva (2010) was used to classify the research. From the point of view of its nature, it is classified as applied, in that it refers to knowledge for practical application aimed at solving specific problems. The qualitative approach considers that there is a dynamic relationship between the real world and the subject, i.e. an inseparable link between the objective world and the subject's subjectivity that cannot be translated into numbers; the quantitative approach means translating opinions and information into numbers in order to classify and analyse them. From the point of view of its objectives, this is a descriptive study, which involved a bibliographical survey, interviews with people who have had practical experience of the problem being researched, the application of questionnaires and systematised observations. In terms of technical procedures, it is a case study, i.e. an in-depth study that allows for broad and detailed knowledge.

4.3. Observation of agricultural activities in the communities

During the months of preparing the land for planting, planting and harvesting, systematised and non-interventionist observations were carried out, according to Mynaio and Gomes (2013); Prodanov (2013). From these day-to-day experiences in the community, it was possible to draw up a specific profile of the practices and methods used in agricultural production in the communities settled by the São Francisco River Transposition Project.

4.4. Social, economic, technological and agro-ecological deterioration

The methodology used to obtain the results was to prepare a questionnaire adapted from Rocha (1997) to be applied to farmers; this methodology was adapted to the context of the Brazilian semi-arid region by Barros (2014). The methodology consisted of surveying and analysing the social, economic, technological and agro-ecological situation at the family identification level. Factors and variables were used to determine the deterioration index (Table 01). Indicators were selected for each variable and scored from 1 to 10 according to the level of deterioration. A higher score represents greater deterioration and a lower score represents a lower level of deterioration.

Table 01. Factors and variables researched

Factor	Variables
Social	Demographics; housing; availability of food; participation in organisations (associations); rural health and enforcement of laws.
Economic	Production; working animals; production animals; marketing,

	credit and income.
Technological	Technology and machinery and verticalisation of production (Rural Industrialisation).
Environmental	Logging; livestock management; erosion; exploitation of native species; soil and water management; use of agrochemicals; rubbish and sewage disposal.

* Indicators were selected for each variable to make up the questionnaire.

4.4.1. Research subjects

Forty-five (45) "indemnified/resettled" families took part in the study. The subjects were selected randomly by drawing lots. After surveying the group to be investigated, those who signed the Informed Consent Form took part in the research, taking into account the ethical standards established by current legislation.

4.4.2. Determining the deterioration index

To tabulate the data, codes were assigned to each item in the questionnaire. The higher the number, the higher the level of deterioration of the factor, and the lower the number, the lower the level of deterioration of the factor.

To determine the deterioration index (y), the straight line equation will be used: $y = ax + b$, where y varies from 0 to 100 (zero to 100 per cent). The minimum x and maximum x' values define the model values a and b, respectively.

The deterioration index was determined from the straight line equation using the maximum and minimum code values and the significant value found in the region, the mode. The deterioration index can vary from zero to 100 per cent.

y - deterioration index (%)

x - modal value found

x'and x'- minimum and maximum values, respectively

a and b - coefficients of the equation of the line

4.4.3. Determining the number of classes and categorisation

To define the number of classes and the subsequent categorisation of the deterioration index, the definition of classes proposed by Sturges (1926) was used, as shown below:

$k = 1 + 3.3 \log 10 (n)$

In which:

k is the number of classes

n is the sample size

The amplitude (A) of each factor analysed was obtained according to the following expression:

A= Vmax - Vmin)

In which:

A is the amplitude of each factor

Vmax is the sum of the maximum values found for the factor

Vmin is the sum of the minimum values found for the factor

The amplitude of the class interval in each factor analysed was determined by the expression:

$h = A/k$

In which:

h is the amplitude of the class interval

k is the number of classes

A is the amplitude of the factor analysed

4.5. Statistical analysis

The data on the level of socio-economic, technological and agro-ecological deterioration was analysed using descriptive statistics, calculating measures of position (mean, median and mode), dispersion (maximum value, minimum value and coefficient of variation). The data was tabulated by grouping together the codes with the highest frequency and repeating them; this highest frequency is called the "mode".

CHAPTER 5

RESULTS AND DISCUSSIONS

5.1. Visits to settled rural communities

As planned in the research project schedule, visits were made to the resettlement sectors in the municipality of São José de Piranhas. Two visits were made to monitor the construction work, the installation of the resettled families and the start of agricultural planting by local residents.

5.1.1. First visit

On 9 September 2015, a technical visit was made to these villas, with the aim of getting to know them and analysing the progress of their works, making it possible to make the following observations:

Vila Produtiva Rural Irapuá I: a village made up of 30 houses, all of which are in the process of being built, with electricity and access roads to the homes. Compared to the others, construction is behind schedule.

Figure 01. Rural Productive Village Irapuá I

Vila Produtiva Rural Irapuá II: a village made up of 20 houses, only one of which is finished, with a sports court under construction. Electricity distributed along the access road to the houses.

Figure 02. Access road and electricity in Vila Produtiva Rural Irapuá II

Vila Produtiva Rural Cacaré: this is considered to be the largest village as it contains 120 houses, all of which have been completed, a sports court that has been partially built and a square that is in the works. Electricity is being installed and several roads give access to the houses.

Figure 03. Houses completed in the Cacaré Rural Productive Village

Vila Produtiva Rural Quixeramobim: the village's construction work has all been completed, with only the electrical installation missing in the 47 houses that make it up. In addition, the water tank for its supply is already positioned at the highest point of the village, supplying water to the homes.

Figure 04. Houses, electricity and access road in Vila Produtiva Rural Quixeramobim

Figure 05. Water tank for supplying the Quixeramobim Rural Productive Village

As a result, it was observed that the Resettlement Programme is relevant for all those who will benefit, as it will guarantee productive development through irrigated agriculture, given the precarious conditions caused by the scarcity of water in the region. The VPRs are undoubtedly an opportunity for economic and social growth in the midst of the difficulties faced in the semi-arid region. In São José de Piranhas-PB, of the four existing villages, only one has been completed, and electricity is in the process of being installed in all of them. Despite this, it is already possible to see from their structures that they have innovative capacities for community living and rural agricultural development.

4.5.1. Second visit

On 24 February 2016, the second visit was made to the Rural Productive Villages (VPRs) in the municipality of São José de Piranhas-PB, which is part of Section III of the Northern Hub of the Project for the Integration of the São Francisco River with the Hydrographic Basins of the Northern Northeast. The main aim of the visit was to analyse the situation of the villages and the productive activities started by farmers in the area. On that occasion, talks were held with some of the farmers who had benefited.

The four existing villages, called Irapuá I, Irapuá II, Cacaré and Quixeramobim, have all been completed. However, only Irapuá I, Irapuá II and Cacaré have had their houses handed over to the owners, who will soon be moving in.

Figure 06. Productive village Irapuá I

Figure 07. Community Centre in Vila Produtiva Irapuá I

Figure 08. Square in Vila Produtiva Cacaré

In some of the homes in Irapuá I and Irapuá II, the beneficiaries are investing in increasing the number of rooms and improving the flooring. In addition, they have also planted corn and beans in the areas around some of the houses.

Figure 09. Investment in improving the floor of a residence in the productive rural village of Irapuá I

In the productive villages, some farmers have already started growing maize and beans in the area around their homes, and it was noted that the farmers are adopting the conventional model of agricultural cultivation. Some farmers are enthusiastic about the possibility of receiving training in the near future to adopt conservationist agricultural practices, while others are incredulous about the possibility of adopting these practices.

Figure 10. Maize and bean plantations in Vila Produtiva Irapuá II

A visit was also made to the primary school that will cater for the children who will live in the productive villages. Due to the high demand provided by its proximity to the villages, the school has been extended to include more classrooms, as well as the benefit of a slab cistern.

Figure 11. Expansion of the school near the Productive Villages

As a result, the Rural Productive Villages (VPRs) in São José de Piranhas-PB will soon be inhabited and have an infrastructure that is consistent with the proposal to improve the quality of life of the farmers who were expropriated by the Integration Project.

5.2. Social diagnosis

The social conditions of the farmers resettled in the rural production villages were diagnosed by analysing the results of the demographic, housing, food availability, organisational, health and law enforcement variables.

The study involved 45 individuals, 55% of whom were male and 45% female. Queiroga et al. (2014), when carrying out socio-economic and environmental studies in the Veneza Settlement, Aparecida-PB, diagnosed that 51% of its population is made up of males and 49% of females.

The owners interviewed were over 25 years old (Figure 12), of which 17.8 per cent were aged between 41 and 45 and the 46 to 50 age group also showed the same percentage. According to Barros (2014), these figures are due to the lack of profitable activities in semi-arid regions, causing young people to migrate, with the majority remaining elderly and children.

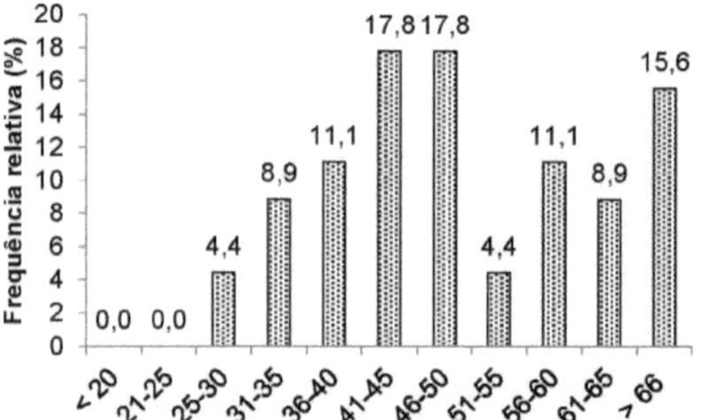

Figure 12. Relative frequency of the age group of the head of the family

This fact was also diagnosed by Santos (2013), who, when researching family farmers in the municipality of Janduís-RN, obtained a majority of interviewees, 38.5 per cent, aged between 41 and 60.

Of all those surveyed, 18.6% have an average family age of between 31 and 35 (Figure 13). As for the total number of family members, 55.6% have 3 to 4 people; 17.8% have 5 to 6 people; 6.7% have 7 to 8 people and 20% have 1 or 2 people.

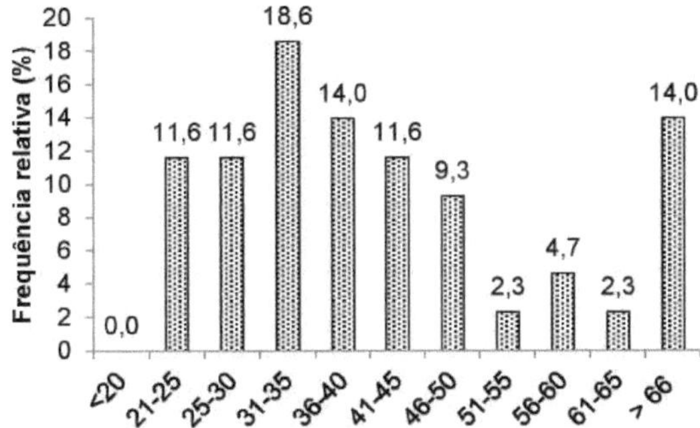

Figure 13. Relative frequency of the average age of the family nucleus

From the results in Figure 14, it can be seen that more than half of the owners surveyed had completed basic education I, comprising 64.4% of the sample, while 20% of them declared themselves illiterate. These figures differ from those obtained by Oliveira et al. (2012) in the municipality of Cachoeira dos Índios, where found that 70% of farmers were illiterate. A high illiteracy rate was also diagnosed by Lacerda et al. (2010) in a survey carried out in the rural community of São Francisco, in the municipality of Conceição-PB, where 42% of those interviewed said they were illiterate.

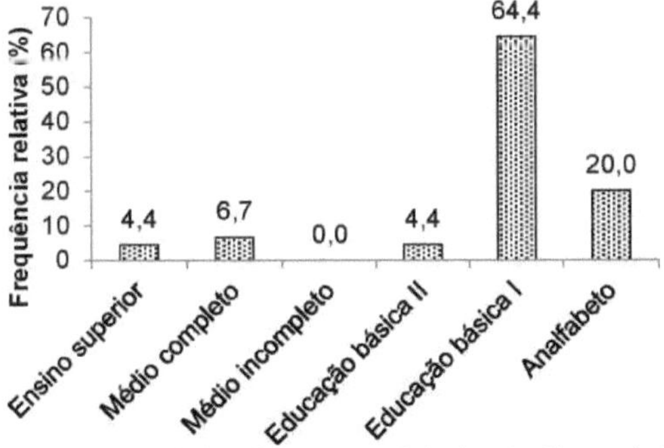

Figure 14. Relative frequency of the head of household's level of education

This data undermines the resettlers' perception of the need to adopt sustainable agricultural practices, as well as pointing to the need to promote public policies aimed at mitigating the low level of schooling.

Ney et al. (2009), when analysing micro-data from the National Household Sample Survey (PNAD) for 2005, stated that the low level of schooling among farmers with less land to plant can compromise the promotion of an agrarian reform policy.

With regard to the type of housing, 97.8% of the dwellings were classified as good masonry and 2.2% as excellent masonry. Similar data was obtained by Oliveira (2013), who, when carrying out research into the use of social water technologies in the semi-arid region of Paraíba, found that 98% of the families interviewed lived in masonry houses. These results are important indicators of a reduction in the proliferation of contaminants found in unhealthy environments.

With regard to the number of rooms in the houses, 97.8 per cent of them have between six and seven rooms and only 2.2 per cent have eight and nine rooms, with an average number of one person per room of 77.8 per cent and two people per room of 22.2 per cent.

With regard to the type of cooker used in households, 84.4% only use gas as a source of energy for domestic activities, 13.3% alternate between using gas and firewood or charcoal and only 2.2% use charcoal or firewood. The high rate of use of gas was due to the recent reintegration of homeowners during the research period, as most of them reported a preference for alternating between the use of gas, firewood and charcoal as an alternative for reducing household costs.

According to the results obtained, all the interviewees have access to drinking water for human consumption, which according to Brito (2013) considerably reduces the number of illnesses in rural areas.

The majority of owners (51.1 per cent) said that their drinking water came from a well; 24.4 per cent said they got it from a water tanker and another 24.4 per cent from the public network.

Unlike these, the results found by Barros (2014) in the Val Paraíso micro-basin, in the semi-arid region of Paraíba, showed that 79.4 per cent of the interviewees carry out some kind of water treatment for consumption and that 75 per cent of them consume water from tube wells.

With regard to the disposal of waste, it was analysed that all the homes visited have septic tanks, contributing to a reduction in the contamination of surface and groundwater. This data is similar to that found by Araújo (2015) in the São Gonçalo Irrigated Perimeter in Sousa-PB, where the majority of those surveyed (76.3%) use septic tanks to dispose of their waste.

As for the disposal of solid waste, it was found that all the homes visited are served by a public collection service, thus contributing to the health of the rural productive villages. Contrary to this result, Queiroga et al. (2014) found that in the Veneza Settlement, in Aparecida-PB, the majority (86.36%) of solid waste is burnt.

When asked how they dispose of empty pesticide packaging, only 11.8% of farmers said they triple wash and collect it from the selling companies; 88.2% said they dispose of it in other ways . This data is worrying because, as Oliveira (2012) emphasises, pesticide packaging waste contains chemical substances capable of modifying the environment and its biota, compromising the natural chain as well as human health.With regard to the number of household appliances and electronics in the homes visited, it was found that all of them have more than three items, with 45.5% of the owners having between five and six items; 27.3% have between three and four items; 20.5% have between seven and nine and 6.8% have more than ten items.The food variable analysed food consumption during the days of the week. As shown in Table 02, the most frequent indicators were the consumption of milk and dairy products, meat, fruit, pulses and vegetables, rice and beans, coffee/tea, corn-derived foods, bread, biscuits and crackers and rapadura/sweets. It can be seen that the interviewees' diet is rich in carbohydrates and proteins, corroborating Abreu (2013).

Table 02. Weekly frequency of food consumed

Item Consumed	**Number of times a week (%)**							
	None	**A**	**Two**	**Three**	**Four**	**Five**	**Six**	**Seven**
Milk and dairy products	15,6	2,2	2,2	6,7	0,0	4,4	0,0	68,9
Meat	0,0	0,0	2,2	20,0	0,0	4,4	2,2	71,1
Fruits	4,4	6,7	20,0	15,6	8,9	0,0	0,0	44,4
Vegetables	8,9	0,0	15,6	13,3	2,2	6,7	0,0	53,3
Potato	22,2	6,7	13,3	24,4	4,4	2,2	0,0	26,7
sweet/maple								
Eggs	22,2	13,3	17,8	15,6	4,4	8,9	0,0	17,8
Pasta	6,8	20,5	36,4	9,1	4,5	2,3	0,0	20,5
(macaroni)								
Rice and/or beans	0,0	4,4	0,0	0,0	0,0	0,0	0,0	95,6
Fish	46,7	35,6	13,3	4,4	0,0	0,0	0,0	0,0
Birds	0,0	0,0	0,0	100,0	0,0	0,0	0,0	0,0
Coffee/tea	2,2	0,0	2,2	0,0	0,0	0,0	0,0	95,6

Derivatives from maize	9,3	14,0	18,6	11,6	2,3	2,3	0,0	41,9
Bread, biscuit, biscuit	4,4	0,0	4,4	11,1	2,2	0,0	0,0	77,8
Rapadura/sweet	18,2	4,5	15,9	11,4	4,5	0,0	0,0	45,5
Flour flour cassava/tapioca	15,6	13,3	24,4	15,6	6,7	0,0	4,4	10,0

As for the participation in organisations variable, the results showed that 91.1% of farmers participate in the local community association and are union members, while 8.9% do not participate in any of these organisations. These data stand out as important indicators for rural development, given that rural organisations provide joint action aimed at improving the socio-environmental conditions of small farmers. According to Santos (2013), family farmers in Janduís-RN also had this characteristic, with 36% of them being members and another 36% being unionised, with the majority participating in both the community association and the Rural Workers' Union.

When asked about pest infestation in the crops grown during the study period, 46.4% of farmers characterised it as high, while only 14.3% characterised it as zero (Figure 15).

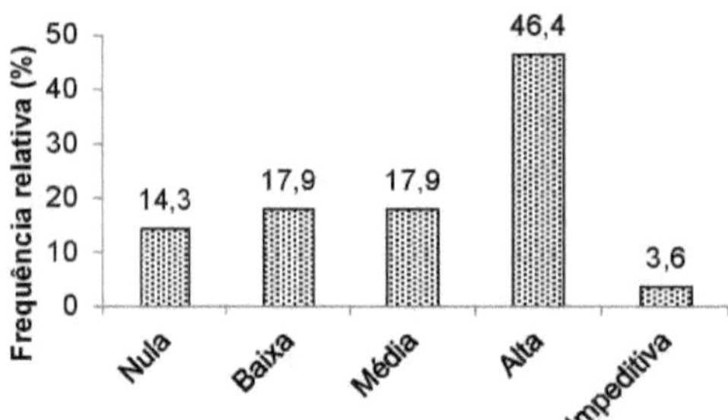

Figure 15. Relative frequency of pest infestation

The high incidence of pests makes it necessary to take measures to combat them in order to maintain production systems. However, the measures adopted are mostly provided by the conventional model of agriculture, based on the uncontrolled use of chemical agents that offensively affect natural resources. This fact was verified in this study, where 69% of the interviewees said that they fight pests using pesticides; only 3.4% fight them using biological control and 27.6% don't use any control techniques.

Similar results were obtained by Melo et al. (2012) in a survey of small producers in the Várzeas de Sousa Irrigated Perimeter, where half of the producers reported using pesticides to control pests and only 17% reported using some kind of biological technique.

The productive villages had a regular level of human health (37.04%). This level characterises an environment with mild relative humidity and little presence of endemic diseases. This is evidence that the quality of life of farmers is improving, as well as pointing to the search for policies that maximise aspects of social wellbeing and combat aspects that prevent human health levels from rising.

None of the households visited were found to have child labour. This indicator is important given the high rate of children and adolescents working in family farming in Brazil. In addition, the absence of child labour contributes to reducing the illiteracy rate in rural areas, given that most of the time children and adolescents drop out of school to help their parents with agricultural activities.

The socio-economic study carried out by Santana et al. (2008), in the community of Pindoba in the municipality of Areia-PB, diagnosed that 75% of the properties have child labour and only 25% of them do not accept this type of work.

The results show that 84.4 per cent of the farmers surveyed work only in subsistence farming, while 15.6 per cent of them alternate between farming and other economic activities in order to generate extra income.

The variation in values (maximum, minimum and modal) for the demographic, housing, health, organisational, food availability and enforcement of laws variables is shown in Figure 16. According to this figure, the organisational and enforcement variables have a modal value close to the maximum value assigned, thus contributing to the increase in the deterioration index of the social factor.

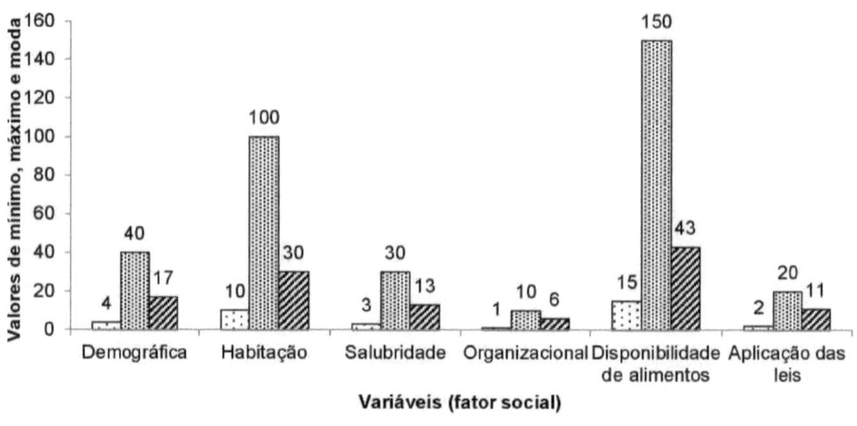

Figura 16. Maximum, minimum and mode values for the social factor variables

To categorise the variables analysed and the level of socioeconomic deterioration, five classes were used, with intervals of twenty units each, according to Abreu (2013). The classes were categorised as very low deterioration, low deterioration, medium deterioration, high deterioration and very high deterioration, as shown in Table 03.

Table 03. Categorisation and class intervals

Classes	Class Range (%)
Very low deterioration	0-20
Low deterioration	20-40
Medium Deterioration	40-60
High deterioration	60-80
Very high deterioration	80-100

According to Figure 17, the line for the food availability variable showed the smallest slope, which indicates that any variation in the significant value will lead to a small variation in the deterioration index. On the other hand, the line for the organisational variable had a steeper slope, given that only one variable was checked, i.e. a small variation in the significant value will lead to a greater variation in the deterioration index.

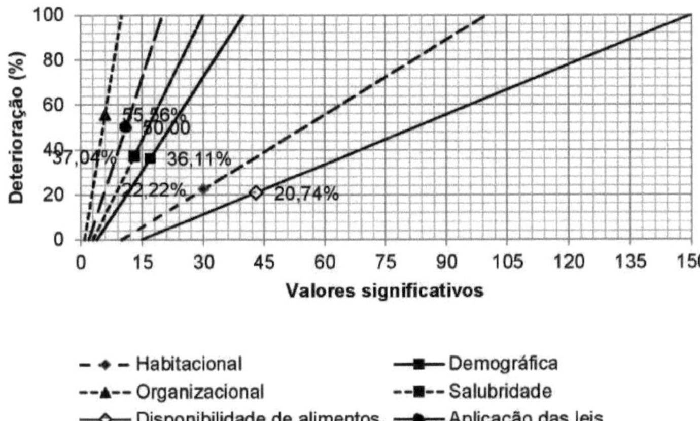

Figure 17. Deterioration for each variable of the social factor

According to Table 02, the food availability, housing, demographic and health variables were classified as low deterioration, while the organisational and enforcement variables were classified as medium deterioration.

The food availability variable showed the lowest deterioration value (20.74%) for the social factor. A similar result was found by Sousa (2010) in the Riacho das Piabas watershed, which showed an overall deterioration index of 29.41% for food consumption on weekdays. The highest deterioration index was for the organisational variable (55.56%), a value close to this was obtained by Pereira and Barbosa (2009) in a study carried out in the watershed of the municipality of São João do Rio do Peixe-PB, where the organisational deterioration index was 53%.

The low rates of deterioration found in the housing, demographic and health variables of 22.22%, 36.11% and 37.04% respectively were lower than those found by Barros (2014) in the Val Paraíso Creek catchment area, where the housing, demographic and health variables were 29.63%, 75% and 55.56% respectively.

The variables of food availability and housing were the ones that contributed most to the low deterioration index of the social factor (26.98%) shown in Figure 18. Ferreira et al. (2008), when analysing the socioeconomic conditions of the São José do Sabugi-PB watershed, diagnosed a low social deterioration index (30.98%). Torres and Vieira (2013), when studying the socioeconomic and environmental conditions of the Córrego dos Pintos watershed in Uberaba-MG, diagnosed a social deterioration index of 39%.

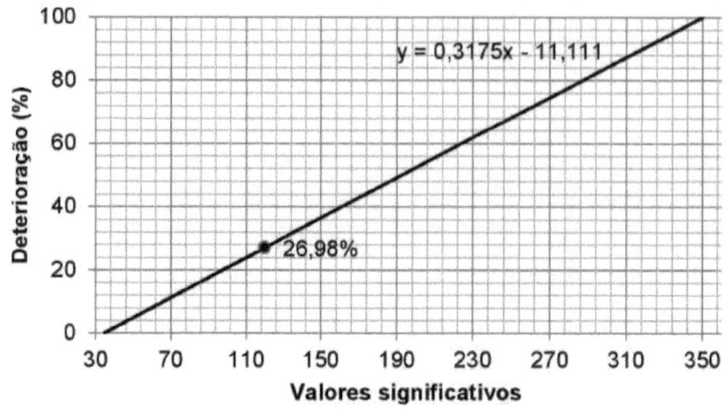

Figure 18. Deterioration of the social factor

According to Table 02, the social factor is classified as low deterioration, although according to the methodology adopted by Rocha (1997) the 26.98% index is well above the tolerable deterioration value of 10%. A result higher than this was obtained by Franco et al. (2005) when they carried out a socio-economic and environmental study of a micro-watershed in the municipality of Boqueirão-PB, where they found a social deterioration index of 62.72%.

5.3. Economic Diagnosis

The main economic activity carried out by the owners of the rural productive villages is farming, particularly maize, beans and poultry. In 2016, these practices were negatively affected by local weather conditions and by the fact that productive plots were not distributed during the favourable planting season, since rainfed agriculture predominates in the region.

According to Figure 19, 48.9% of the farmers did not produce. The results corresponding to the below average and average production indicators, 40% and 11.1% respectively, were represented by farmers who grew maize and beans in the areas around their homes or in other private areas.

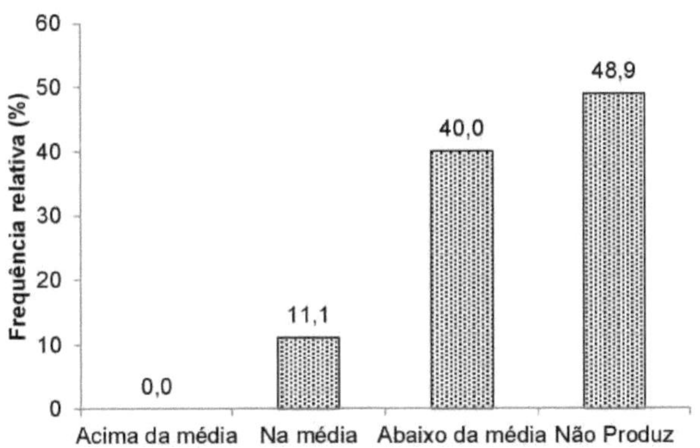

Figure 19. Relative frequency of average agricultural productivity

The socio-economic study carried out by Abreu (2013) in the Riacho Fundo micro-basin, in the municipality of Cabaceiras-PB, also diagnosed a high rate (86%) of families that do not carry out any kind of productive activity.

With regard to afforestation, it was found that all the properties studied had a native vegetation index of more than 20% of the area. This is due to the fact that the resources available in the areas used for farming were not exploited during the study period. However, if there is no public investment in training the owners of the productive villages to develop conservationist agricultural practices, these areas could become the scene of the degradation processes seen in semi-arid regions.

The majority of the resettled farmers surveyed (88.9%) do not have planted pastures; 8.9% have conserved pasture without a strategic food reserve and only 2.2% buy bulk feed to feed their livestock. In contrast to this, the result obtained by Pisani et al. (2011) in the Rio das Pedras sub-basin, in the municipality of Itatinga-SP, was that 90% of the owners reported having planted pastures in a good state of conservation.

According to Figure 20, 86.7% of the farmers interviewed do not own any type of animal for rural work; 8.9% own one work animal and only 4.4% own two types of work animals. The low number of working animals, as well as production animals, is associated with the fact that at the time of the study the owners had not yet taken possession of their productive plots so that they could carry out farming activities.

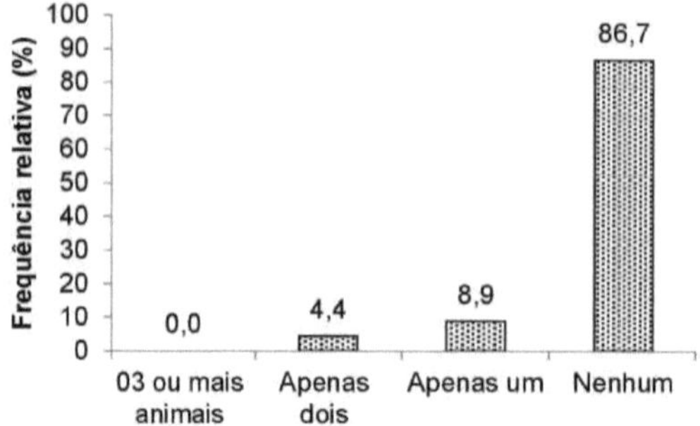

Figure 20. Relative frequency of working animals

As for the livestock variable (Figure 21), 42.2 per cent of the resettled people have only one type of livestock, while 40 per cent have no livestock at all. The most common type of production in the households surveyed is poultry, given its low cost and the availability of resources for its maintenance. These figures corroborate those found by Silva and Costa (2012) in the Recôncavo Baiano, where 96% of family farmers declared that they kept animals, 87% of whom kept poultry.

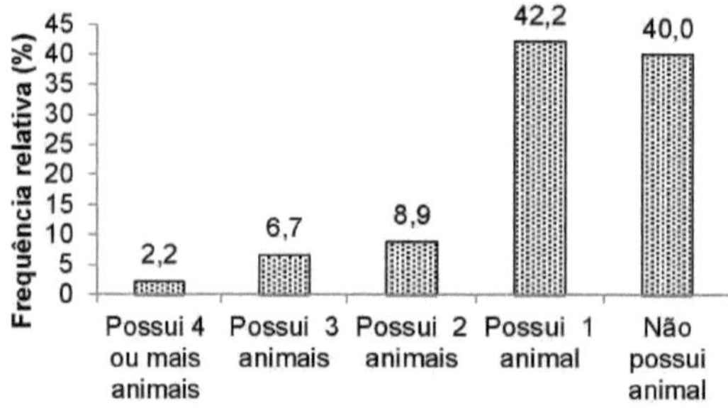

Figure 21. Relative frequency of production animals

When asked about the commercialisation of agricultural products, 97.8% of the resettled people said they didn't commercialise, while only 2.2% sold directly to consumers. With regard to the marketing of livestock products, it was found that the majority of those surveyed (82.2%) do not market, and that only 8.9% sell directly to consumers; 6.7% sell to meatpackers and 2.2% sell to traders. It was also found that farmers do not sell forestry products.

Nardini (2013), when analysing the profile of rural producers for the agricultural, livestock and forestry production variable in the Água-Fria catchment area in the municipality of Bofete-SP, found that the majority of larger rural producers sell their products directly to agro-industries, while small producers sell their produce to intermediaries, warehouses or not at all.

According to the results, only 17.8% of the resettled farmers have access to agricultural credit through official banks, while 82.2% of them do not have access to any form of agricultural credit. Melo et al. (2012) found that in the Várzeas de Sousa Irrigated Perimeter, 37% of small farmers have access to credit to finance their production and 63% of them have not been able to access it, the main reason being the difficulty encountered in receiving this benefit.

When analysing the approximate gross income of the properties surveyed, it can be seen that most of them (91.1%) have an income of up to half the minimum wage and that only 8.9% have an income of up to one minimum wage, which is due to the low rate of agricultural production found in the rural productive villages. It was also found that the majority of landowners (97.8%) have other types of income such as the Bolsa Família, retirement and federal government aid for those affected by the São Francisco river transposition works, while 2.2% declared that they had no other types of income. Cruz et al. (2010), when analysing the monthly income of farmers living in the Nova Vida settlement in the municipality of Mogeiro-PB, found an average income of R$499.00, with 33.3% receiving the Bolsa Família and 14.28% receiving a pension.

Figure 22 shows the total monthly income of the farmers interviewed. It shows that most of them (46.7%) have an income of between one and two minimum wages. In contrast to these, the results found by Gonzaga et al. (2015) in their socioeconomic diagnosis of family farmers in Acre were 53% of those who said they had a monthly income of between one and two minimum wages.

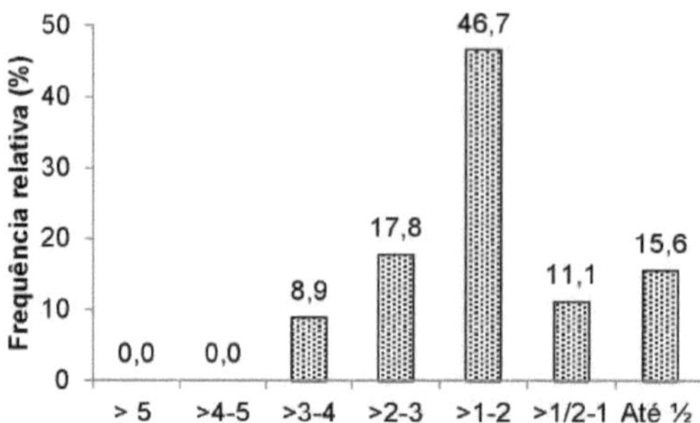

Figure 22. Relative frequency of total monthly income

According to Figure 23, the production and livestock variables showed modal values close to the maximum values assigned and the livestock variable showed a modal value equal to the maximum value assigned, thus contributing to the high rate of economic deterioration in the area studied.

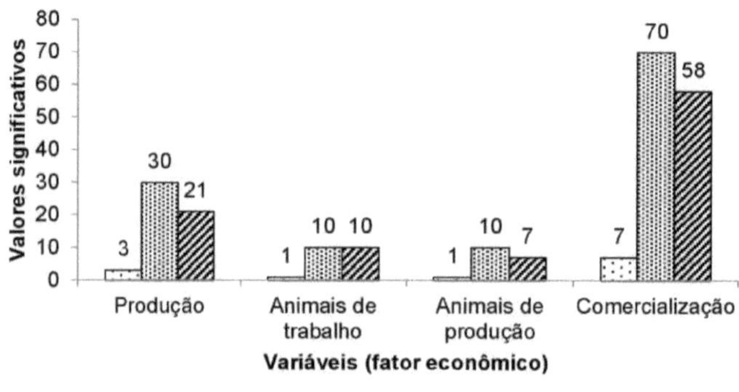

Figure 23. Maximum, minimum and mode values for the economic factor variables

An analysis of the deterioration indices for each economic factor variable (Figure 24) showed that the production and livestock variables were classified as highly deteriorated, and the marketing and livestock variables were classified as very highly deteriorated.

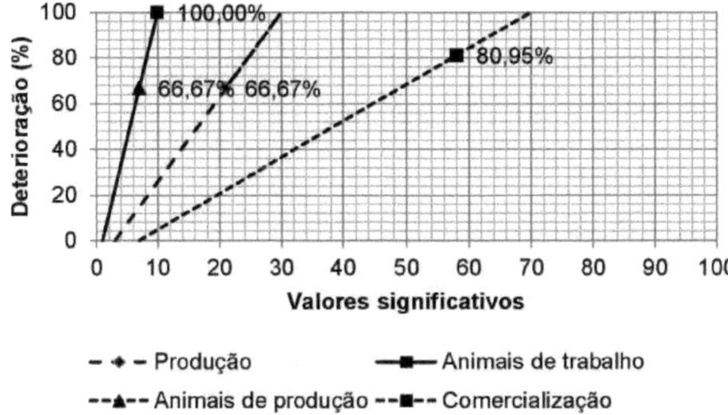

Figure 24. Deterioration for each economic factor variable

The working animals variable had the highest deterioration index (100%) for the economic factor. This was followed by the commercialisation variable with a deterioration index of 80.95% and the production and livestock variables with the same deterioration value (66.67%). These results are due to the high frequency of the indicators' maximum values, which generated high modal values.

In the diagnosis carried out by Franco et al. (2005) in the Boqueirão-PB watershed, it was found that the production and working animal variables had the highest deterioration index (100%) for the economic factor, while the production animal and commercialisation variables had a deterioration index of 44.4% and 80.87%, respectively.

The deterioration index for the economic factor was 77.78 per cent (Figure 25). According to Table 02, this value is characterised as high deterioration, and is well above the value of 10% stipulated by Rocha (1997).

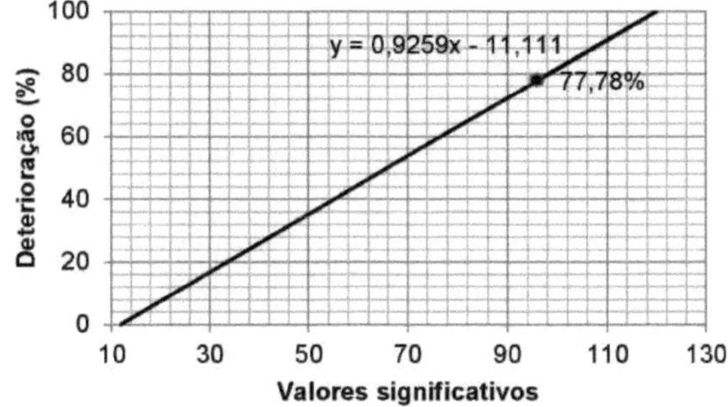

Figure 25. Deterioration of the economic factor

This result was higher than that found by Ferreira et al. (2008) in the municipality of São João do Sabugi-PB, where they found a deterioration rate of 63.33%. Abreu (2013) also diagnosed a deterioration rate (51.21%) in the municipality of Cabaceiras-PB that was lower than that found in this study.

5.4. Technological diagnosis

The rural production villages surveyed have less than 20 ha and 50% utilisation, thus being characterised as smallholdings. Similar results were found by Souza and Montenegro (2011) in the community of Sítio Jardim, in the municipality of Areia-PB, where the majority of family producers own between 1 and 8 ha of land. According to these authors, the area of the property is a limiting factor for production, given that the smaller the size available for planting, the less diversification there will be and the greater the negative impacts on the profitability of the properties.

As for the type of tenure, all the owners fell within the indicator of being resettled by the transposition project. Neto et al. (2008) analysed the agro-ecosystems and environmental and socio-economic conditions in the rural area of Vitória do Mearim-MA and found that 33.33% of farmers owned their land and 66.67% were occupiers or tenants.

With regard to the use of pesticides (Figure 26), it was found that 45.12% of producers use them based on empirical advice. These results are worrying, because as Silva et al. (2015) state, pesticides used in agriculture can be considered the main culprits in contaminating natural resources, causing negative effects on human health.

According to Mendes et al. (2014), pesticides are the resources most used by producers to try to compensate for production problems caused by soil degradation and pest infestation.

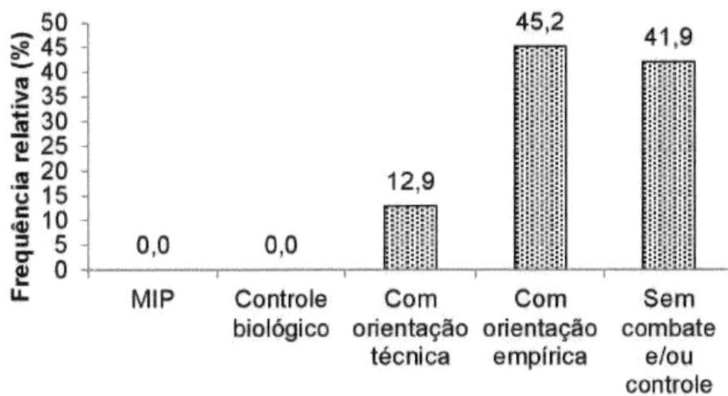

Figure 26. Relative frequency of pesticide use

When asked about fertilisation and/or liming to correct and nourish the soil, 90.3% of the interviewees said they did not adopt these practices and only 9.7% carried out chemical fertilisation through technical guidance. This fact can be considered one of the negative points that resulted in the low production rate found in the productive villages, because as Barros (2014) states, appropriate fertilisation and liming practices can provide good profitability in agriculture.

Similar data was obtained by Santana et al. (2008) in the socio-economic diagnosis of family farmers in the municipality of Areia-PB, where it was found that 95% of them do not use any fertilisation technique and only 5% use chemical and organic fertilisation.

Regarding the type of tool used to carry out farming activities, it was found that 85% of farmers use only hand tools, 12.5% use only mechanical tools and 2.5% use both types. Similar results were found by Araújo (2015) in the semi-arid region of Paraíba, where 88% of farmers use only hand tools, 4% mechanical and 8% both.

With regard to the type of transport used to transport production and to get around, 42.2% of the interviewees use a motorbike, 33.3% use a car, 4.4% use a bicycle and 20% don't have any type of transport.

With regard to land preparation practices for planting, in terms of the direction of the beds, 59.4% of the interviewees prepared the soil in a contour and 40.6% in favour of the slope. This data shows that most of the farmers surveyed use contour planting, thus favouring the containment of nutrients in the soil and avoiding erosion.

As for animal feed reserves, 91.1 per cent said they didn't have any, while only

8.9 per cent said they had energy crops or bulk feed for cattle and poultry.

When asked about soil conservation practices, 66.7% of the farmers said that they didn't carry out any type of practice and 33.3% said that they adopted conservation management practices, mainly the contour planting mentioned above. In contrast, the results found by Pisani et al. (2001) in the Pedras River sub-basin in Itatinga-SP were that 66.5% of farmers carried out some kind of soil conservation practice and 33.5% did not adopt any conservation practice.

The use of irrigation for production is not yet a reality in the rural production villages, given the unfinished work on the transposition of the São Francisco River. Even so, the resettled people's expectations about the possibilities for developing production systems that will be provided by the transposition were perceptible.

According to Castro (2012), irrigation is a technology of great importance for agricultural production in the northeastern semi-arid region. This is evident from the fact that this region has low rainfall rates that limit the development of agricultural activities.

According to Figure 27, 82.2% of those interviewed do not receive technical assistance. A higher result was found by Souza and Montenegro (2011) in the community of Sítio Jardim, Areia-PB, where 96% of family farmers do not receive technical assistance. According to these authors, the lack of technical assistance hinders rural development, given that family farmers need to expand their knowledge in order to remain in the countryside.

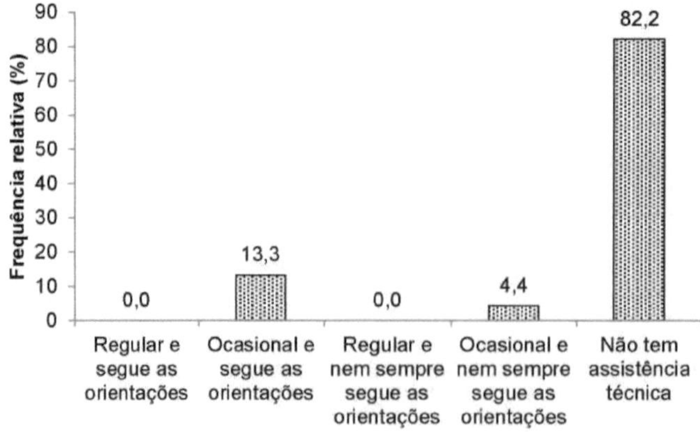

Figure 27. Relative frequency of receiving technical assistance

With regard to the adoption of agricultural practices to exploit the land, the most frequent indicators were deforestation and burning. This diagnosis points to the need for farmers to adopt sustainable agricultural practices, because as Brasileiro (2009) emphasises, the removal of native vegetation cover is one of the first indicators of the degradation and desertification processes witnessed in semi-arid regions.

As for the existence of water collection techniques, 19.5 per cent of the properties have no water collection techniques and 80.5 per cent collect water through tube wells. It is important to emphasise that the existence of tube wells is at the level of the productive village, i.e. most villages have a well to supply all the homes, and not at the level of individual properties.

When asked about the way in which their livestock is farmed, 83.3% of the owners said that they adopt the semi-extensive system, where animals are rounded up in the afternoon for supplementary feeding; 16.7% adopt the intensive system, where animals are confined in paddocks with 80% of their feed in troughs, and none of them adopt the extensive system. Sousa (2010) also diagnosed a high rate (54.78%) of producers adopting the semi-extensive system of rearing at the headwaters of the Riacho das Piabas.

It was found that only 26.7% of the owners have some kind of machine to help with their farming activities, while 73.3% don't have any. Among the machines used, the costal sprayer, cart and forage harvester stand out.

With regard to making some kind of handicraft, the data shows that 93.3% of the residents interviewed do not make any kind of handicraft. This is a negative fact, because as Ferreira et al. (2008) state, the lack of exploitation in handicrafts reduces the possibility of extra income for the owners. With regard to adding value by processing raw materials (Figure 28), it was found that 95.6% do not carry out any type of processing. Similar results were found by Barros (2014) in the Riacho Val Paraíso watershed, where 82.4% of those surveyed did not carry out any type of processing.

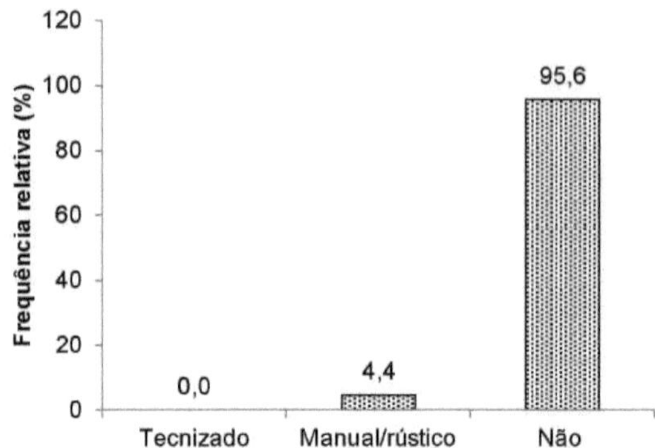

Figure 28. Relative frequency of adding value by processing raw materials on the farm

The variation in the values comprising the technology and rural industrialisation variables, referring to the social factor, are shown in Figure 29. It can be seen that the technology variable had a modal value close to the maximum value assigned and the rural industrialisation variable had a modal value equal to the maximum value assigned, contributing to an increase in the technological deterioration index in the area studied.

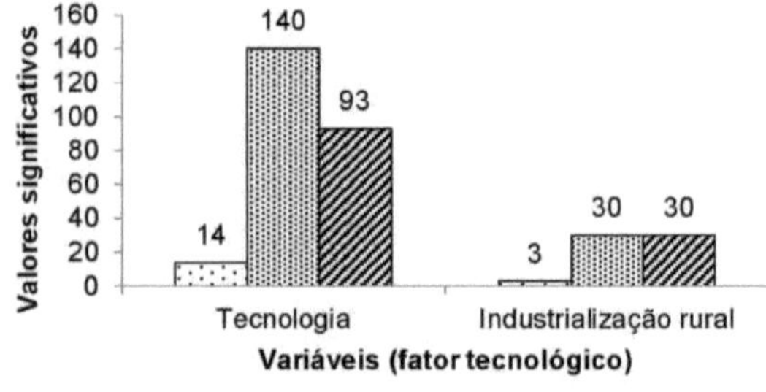

Figure 29. Maximum, minimum and mode values for the technological factor variables

According to Figure 30, rural industrialisation showed the highest deterioration value (100%) and the technology variable showed a of 62.70%, being classified as very high deterioration and high deterioration respectively.

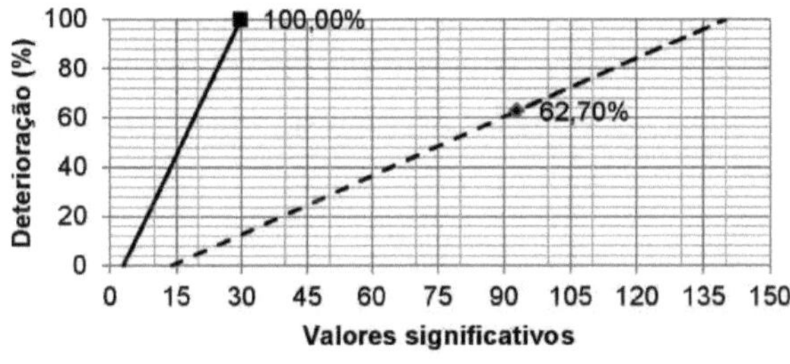

Figure 30. Deterioration for each variable of the technological factor

The indicators that most contributed to the high rate of technological deterioration were the lack of conservation farming practices, irrigation and technical assistance. With regard to the rural industrialisation variable, the indicators that most contributed to its high rate of deterioration were the absence of craft activities and the processing of raw materials.

When investigating the deterioration of the technological factor in a watershed located in the municipality of São João do Rio do Peixe, Pereira and Barbosa (2009) identified a deterioration of 100% for the technological variable and 60.2% for the rural industrialisation variable.

The deterioration index for the technological factor was 69.28 per cent (Figure 31). According to Table 02, it is classified as highly deteriorated and is well above the maximum value assigned by Rocha (1997). This indicates that more than half of the area studied is deteriorated.

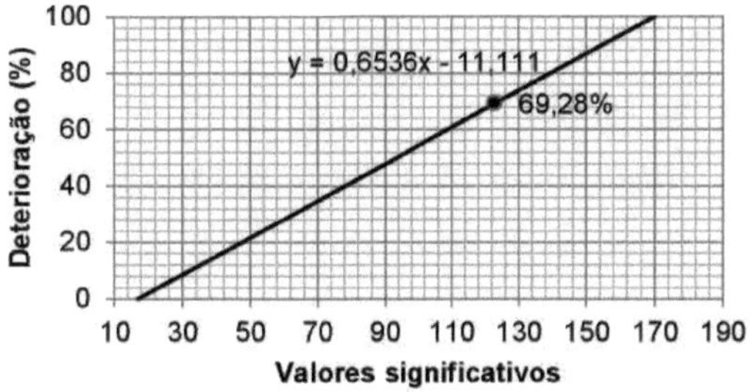

Figure 31. Deterioration of the technological factor

The technological deterioration diagnosed by Silva and Mattos (2013) in the Riacho do Poço de Serra watershed, in the municipality of Currais Novos-RN, was

82.30 per cent. A similar index was found by Alves and Alves (2012) in the socio-economic diagnosis of rural settlements in the state of Paraíba, which showed a technological deterioration of 78.21%.

5.5. Socio-economic diagnosis

The socio-economic deterioration of the area studied was 47.74% (Figure 32). This index is classified as medium deterioration and is well above the maximum value (10%) tolerated by Rocha (1997). Similar results were obtained by Silva and Mattos (2013) in the Riacho do Poço de Serra watershed in Currais Novos-RN, where they diagnosed a socioeconomic deterioration index of 41%. Torres and Vieira (2013), when carrying out a socio-economic and environmental study on a tributary of the River Uberaba, in Uberaba-MG, diagnosed a socio-economic deterioration index of 45%.

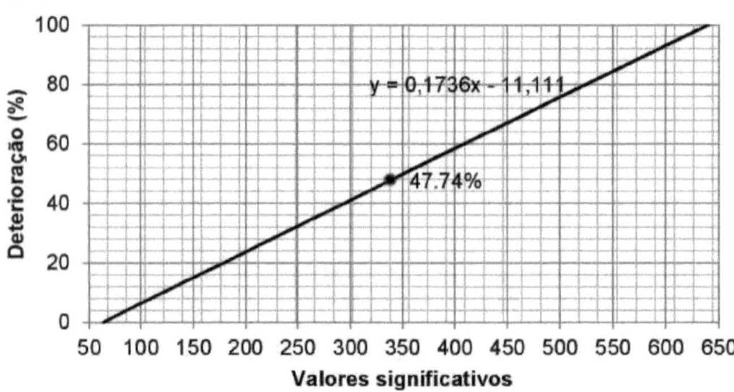

Figure 32. Deterioration of the socio-economic factor

The socio-economic deterioration index found in this research was higher than that found by Barros et al. (2014), in the socio-economic diagnosis of a micro-watershed in the Paraíba hinterland, where they obtained a socio-economic deterioration index of 59.83%.

5.5. Environmental Diagnosis

The environmental conditions of the area studied were analysed using 19 environmental pollution indicators. As shown in Table 04, the results of the modal, maximum and minimum values were 21, 19 and 57 respectively.

Table 04. Diagnostic results and critical unit of environmental deterioration

Environmental Diagnosis				
Code	**Indicators**	**Significant Values**		
		Fashion	**Minimum**	**Maximum**
1.1	Stocking up on pesticides	1	1	3
1.2.	Pesticide packaging depots	1	1	3
1.3.	Places to wash pesticide application implements	1	1	3
1.4.	Application of pesticides	1	1	3
1.5.	Waste bins (urban and rural waste)	1	1	3
1.6.	Logging	1	1	3
1.7.	Wild animal breeding	1	1	3
1.8.	Hunting for sale	1	1	3
1.9.	Pigsties	1	1	3
1.10.	Aviaries/stables	1	1	3
1.11.	Slaughterhouses	1	1	3
1.12.	Marked erosion	1	1	3
1.13.	Open sewers	1	1	3
1.14.	Burning	3	1	3
1.15.	Pumps for pumping water out of rivers/weirs	1	1	3
1.16.	Whey	1	1	3
1.17.	Using insecticides with your hands	1	1	3
1.18.	Exploitation of native species	1	1	3
1.19.	Water use for irrigation	1	1	3
	Total environmental factor	21	19	57
	Critical deterioration unit (y)		5,26%	

It can be seen that most of the indicators had a modal value of 1, indicating low anthropogenic pressure on the natural resources available in the region. Table 05 shows the percentages of values 1, 2 and 3 assigned to each indicator. Value 1 refers to the alternative "does not exist", value 2 to the alternative "exists with technical-scientific guidance" and value 3 to the alternative "exists without technical-scientific guidance". In this sense, it can be seen that indicators 1.1, 1.4 and 1.14 contributed the most to the environmental deterioration found in the area surveyed. On the other hand, indicators 1.3, 1.5, 1.7, 1.8, 1.11, 1.12, 1.13 and 1.15 indicated that there were no negative environmental impacts.

Table 05. Relative frequency of codes 1, 2 and 3

Environmental Diagnosis			
Indicators	Value 1 (%)	Value 2 (%)	Value 3 (%)
1.1. Stocking up on pesticides	95,56	22,22	22,22

1.2.	Pesticide packaging depots	93,33	2,22	4,44
1.3.	Places to wash pesticide application implements	97,78	2,22	0,00
1.4.	Application of pesticides	66,67	11 ,11	22,22
1.5.	Waste bins (urban and rural waste)	100,00	0 , 00	0,00
1.6.	Logging	97,78	0,00	2,22
1.7.	Wild animal breeding	100,00	0,00	0,00
1.8.	Hunting for sale	100,00	0,00	0,00
1.9.	Pigsties	84,55	0,00	15,56
1.10.	Aviaries/stables	88,89	0,00	11,11
1.11.	Slaughterhouses	100,00	0,00	0,00
1.12.	Marked erosion	100,00	0,00	0,00
1.13.	Open sewers	100,00	0,00	0,00
1.14.	Burning	53,33	0,00	46,67
1.15.	Pumps for pumping water out of rivers/weirs	100,00	0,00	0,00
1.16.	Whey	95,56	0,00	4,44
1.17.	Using insecticides with your hands	93,33	0,00	6,67
1.18.	Exploitation of native species	95,56	0,00	4,44
1.19.	Water use for irrigation	97,78	0,00	2,22

The environmental deterioration index found was 5.26% (Figure 33), characterised as very low deterioration and lower than the maximum value of 10% stipulated by Rocha (1997). A similar result was diagnosed by Franco et al. (2005) in the micro-basin of the Epitácio Pessoa Reservoir, Boqueirão-PB, where they found an environmental deterioration of 9.09%.

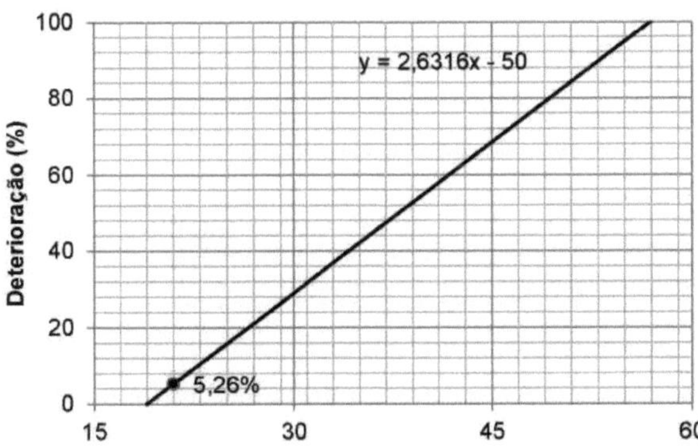

Figure 33. Environmental deterioration

Unlike this, the environmental diagnosis carried out by Souza (2010) in the springs of the Riacho das Piabas-PB identified an environmental deterioration index of 42.86%. Pereira and Barbosa (2009), when assessing the level of environmental deterioration of a micro-basin in the Paraíba hinterland, found an environmental deterioration of 25.9%. Barros and Chaves (2014), when analysing the environmental conditions of the Val Paraíso sub-basin, diagnosed an environmental deterioration index of 56.25%.

CHAPTER 6

FINAL CONSIDERATIONS

Resettled farmers need to adopt sustainable agricultural practices, given that most of them use the conventional model to maintain their production systems.

The practice of deforestation and burning was reported as the main way of preparing the soil for planting. This has negative effects on the soil, contributing to its infertility and consequent loss of productivity. Furthermore, the development of these practices is considered to be the main cause of the high rates of degradation and desertification in semi-arid regions.

With regard to soil management, there was also a predominant lack of fertilisation and conservation practices, which further explains the low productivity index found in the productive villages. Other indicators that favoured the increase in this index include the lack of technical assistance and irrigation. These are important indicators given the low economic and water conditions that limit the development of agricultural activities in the semi-arid region.

The high occurrence of pests that affect crops leads to the systematic use of pesticides, negatively influencing natural resources and the living beings that depend on them.

The levels of social and environmental deterioration diagnosed in the area studied are relevant to the reality witnessed in most semi-arid regions. The study showed that the resettled people have good social conditions, provided by the infrastructure of the productive villages, and there was also little anthropogenic action on Caatinga resources.

On the other hand, the high rates of deterioration in the economic, technological and socio-economic factors indicate the challenges to be faced if farmers living in the productive villages are to become aware of and adopt sustainable agricultural practices, and also point to the need to adopt public policies aimed at mitigating the indicators that favour the deterioration of these factors.

REFERENCES

ABREU, B. S. Local socioeconomics as an index of happiness and environmental perception: a case study in the Ribeira District - Cabaceiras (PB). 2013. 193f. Thesis (Doctorate in Natural Resources) - Federal University of Campina Grande, Campina Grande - PB, 2013.

ALTIERI, M. An agroecological look at industrial agriculture. In:. Agroecology: scientific bases for sustainable agriculture. 3ed. São Paulo, Rio de Janeiro: Expressão Popular, AS-PTA, 2012. p. 20-100.

. Introduction. In:. Agroecology: scientific bases for sustainable agriculture. 3ed. São Paulo, Rio de Janeiro: Expressão Popular, AS-PTA, 2012. p. 15-19.

. Conceptual and methodological bases of agroecology. In:. Agroecology: scientific bases for sustainable agriculture. 3ed. São Paulo, Rio de Janeiro: Expressão Popular, AS-PTA, 2012.p. 101-218.

ALVE, A. R.; ALVES, J. B. Risks and vulnerabilities in rural settlements in the state of Paraíba. Revista Geonorte, v. 2, n. 5, p. 112-1132, 2012.

ALVES, A. P. Coexistence with the Brazilian semi-arid region. In: CONTI, I. L.; SCHROEDER, E. O. (Orgs). Strategies for coexistence with the Brazilian semi-arid region. Brasília: Editora IABS, 2013. p. 35-44.

ANGELOTTI, F.; JÚNIOR, P. I. F.; SÁ, I. B. de. Climate change in the Brazilian semi-arid region: mitigation and adaptation measures. Revista Brasileira de Geografia Física, v.4, n.6, p. 1097-1111. 2011.

ARAUJO, J. M.; ARRUDA, D. B. Sustainability practices in the semi-arid Northeast: the right to sustainable economic development. Veredas do Direito, v.8, n.16, p. 235-260, Jul./Dec. 2011.

ARAÚJO, J. T. de. Water footprint and socioeconomic, technological and environmental conditions of the communities surrounding the São Gonçalo permanent preservation area, Sousa-Paraíba. 2015. 104 f. Course Conclusion Work (Degree in Biological Sciences) -

Federal University of Campina Grande, Teacher Training Centre, Cajazeiras-PB, 2015.

ARAÚJO, I. P. de; LIMA, J. R.; MENDONÇA, I. F. C. Use of degradation and degradation of natural resources in the Brazilian semi-arid region: a study in the Farinha river watershed, Paraíba, Brazil. Caminhos de Geografia, v.12, n.39, p. 255-270, sep. 2011.

ARAÚJO, S. M. S. de. The semi-arid region of north-eastern Brazil: environmental issues and possibilities for the sustainable use of resources. Rios Eletrônica, n. 5, p. 89-98, Dec. 2011.

ASSIS, R. L. de. Sustainable rural development in brazil: perspectives from the integration of public and private actions based on agroecology. Economia Aplicada, v.10, n.1, p. 75-89, jan./mar. 2006.
ASSAD, M. L. L.; ALMEIDA, J. Agriculture and sustainability: context, challenges and scenarios. Ciência & Ambiente, n.29, p. 15-30. 2004.

BARROS, J. D. de S. Carbon and nitrogen stocks in vertisols and socioeconomic and environmental conditions in the Riacho Val Paraíso watershed (PB). 2014. 152 f. Thesis (Doctorate in Natural Resources) - Federal University of Campina Grande, Campina Grande - PB, 2014.

BARROS, J. D. de S.; SILVA, M. de F. P. da. Agroecology and agroecological practices as alternatives to the hegemonic model of agricultural production. Revista Sociedade e Desenvolvimento Rural, v.4, n.2, p. 89-103, 2010.

_______ . Methodology of Study and Scientific Research. João Pessoa- PB: Sal da Terra, 2010.

BARROS, J. D. de S. et al. Riacho Val Paraíso watershed: socioeconomic analysis. Holos, v.4, n.30, p. 34-46. 2014.

BARROS, J. D. de S.; CHAVES, L. H. G. Environmental analysis of the Val Paraíso Creek sub-basin. Brazilian Geographical Journal: Geosciences and Humanities research médium, v.5, n.1, p. 105-115, jan./jun. 2014.

BAPTISTA, N. de Q.; CAMPOS, C. H. Coexistence with the semi-arid region and its potential. In: CONTI, I. L.; SCHROEDER, E. O. (Orgs). Coexistence

with the Brazilian Semi-Arid: autonomy and social protagonism. Brasília: Editora IABS, 2013. p. 51-58.

BRASILEIRO, R. S. Sustainable development alternatives in the northeastern semi-arid region: from degradation to conservation. Scientia Plena, v.5, n.5, p. 1-12, May 2009.

BRITO, P. N. de F. Quality of water supply in rural communities of the Várzea do Baixo Rio Amazonas. 2013. 49 f. Course Conclusion Paper (Graduation) - Federal University of Amapá Foundation, Pro-Rectory of Undergraduate Education, Environmental Sciences Course, Macapá-AP, 2013.

CAPORAL, F. R. Agroecology: a new science to support the transition to more sustainable agriculture. In: (Org). Agroecology: a science from the field of complexity. Brasília: 2009, p. 9-64.

CAPORAL, F. R.; COSTABEBER, J. A. Agroecology and sustainable rural development: perspectives for a new rural extension. Agroecologia e Desenvolvimento Rural Sustentável, v.1, n.1, p. 16-37, jan./mar. 2000.

CASTRO, C. N. de. Agriculture in northeast Brazil: opportunities and limitations to development. Brasília, Rio de Janeiro: Ipea, 2012. 43 p.

CONTI, I. L.; PONTEL, E. Paradigmatic transition in living with the semi-arid region. In: CONTI, I. L.; SCHROEDER, E. O. (Orgs). Coexistence with the Brazilian Semi-Arid: autonomy and social protagonism. Brasília: Editora IABS, 2013. P. 21-30.

CRUZ, R. N. da. et al. Socioeconomic diagnosis of sesame cultivation in the Nova Vida Settlement, PB. In: Congresso Brasileiro de Mamona, 4 and Simpósio Internacional de Oleaginos Energéticas, 1, 2010, João Pessoa- PB. Social Inclusion and Energy: Proceedings... Campina Grande: Embrapa Algodão, 2010. p. 363-368.

DUARTE, R. G. et al. Environmental education in coexistence with the semi-arid region: actions developed by the Ceará state education department. Revista de Gestão Ambiental e Sustentabilidade, v.4, n.1, p. 17-29, jan./abr. 2015.

DIAS, R. Sustainable development as a new paradigm. In:. Environmental Management: social responsibility and sustainability. São Paulo: Atlas, 2009. p. 30 43.

DUARTE, S. M.; BARBOSA, M. P. Study of natural resources and their potential in the semi-arid region, state of Paraíba. Engenharia Ambiental, v.6, n.3, p. 168-189, Sep./Dec. 2009.

FEIDEN, A. Agroecology: introduction and concepts. In: AQUINO, A. M. de; ASSIS, R. L. de (Orgs). Agroecology: principles and techniques for sustainable organic agriculture. Brasília: Embrapa Informação Tecnológica, 2005. p. 29-70.

FERNÁNDEZ, X. S.; GARCIA, D. D. Sustainable rural development: an agroecological perspective. Agroecologia e Desenvolvimento Rural Sustentável, v.2, n.2, p. 17-26, apr./jun. 2001.

FERREIRA, A. C. et al. Socioeconomic diagnosis of the municipality of São José do Sabugi, PB. Revista educação Agrícola Superior, v. 23, n. 1, p.101-104, 2008.

FRANCO, E. S.; LIRA, V. M. de: PORDEUS, R. V.; LIMA, V. L. A. de; DANTAS NETO,
J.; AZEVDO, C A. V. de. Socio-economic and environmental diagnosis of a micro-watershed in the

Municipality of Boqueirão - PB. Engenharia Ambiental, v. 2, n. 1, p. 100-114, 2005.

GONZAGA, D. S. de O. M.; PERES, R. T.; SILVA, F. de A. C. Socio-economic characterisation of family farmers, suppliers of raw material for fruit agro-industries in Acre and on the border of Rondônia. In: Congress of the Brazilian Society of Economics, Administration and Rural Sociology, 53, 2005, João Pessoa-PB. Proceedings... João Pessoa-PB: SOBER, 2015

GUEDES, Z. M.; MARTINS, J. C. de V. Agroecologia e gênero: uma perspectiva socioambiental no Assentamento Mulunguzinho em Mossorró- RN. Revista Verde de Agroecologia e Desenvolvimento Sustentável, v5, n.1, p. 66-76, jan./mar. 2001.

JESUS, E. L. de. Different approaches to non-conventional agriculture: history and philosophy. In: AQUINO, A. M. de; ASSIS, R. L. Agroecologia:

principles and techniques for sustainable organic agriculture. Brasília-DF: Embrapa Technological Information, 2005. P. 21-48.

LACERDA, C. de S.; CÂNDIDO, G. A. Models of sustainability indicators for water resources management. In: LIRA, W. S.; CÂNDIDO, G. A. (Orgs). Sustainable management of natural resources: a participatory approach. Campina Grande: EDUEPB, 2013. P 13-30.

LACERDA, M. B. S et al. Socioeconomic diagnosis of farmers and the impact of the bolsa família programme, Conceição - Paraíba - Brazil. Revista Holos, ano 26, v. 1, p. 41-50, 2010.

LIRA, M. G. da C.; OLIVEIRA, B. R. B. de. The process of transferring and adapting technology and knowledge: the case of EMBRAPA Semiárido. Revista Semiárido de Visu, v.2, n.2, p. 274-284. 2012.

MALVEZZI, R. The Brazilian semi-arid region. In:. Semi-arid: a holistic vision. Brasília: Confea, 2007. p. 9-19.

. The prospects for coexistence with the semi-arid region. In:. Semi-arid: a holistic vision. Brasília: Confea, 2007. p. 105-129.

MENDES, E. do N. et al. The use of pesticides by farmers in the municipality of Tianguá-Ce. Agropecuária Científica no Semi-Árido, v.10, n.1, p. 07-13, jan./mar. 2014

MELO, W. F. de. An analysis of the agricultural production chain in the Várzeas of Sousa-PB: a study of small producers. Revista Verde, v. 7, n. 3, p. 102-108, jul./set. 2012

MINAYO, M. C. de S.; GOMES, S. F. D. R. Pesquisa social: teoria, método e criatividade. 33. ed. Petrópolis - RJ: Vozes, 2013.

MOREIRA, R. M.; CARMO, M. S. do. Agroecology in the construction of sustainable rural development. Revista Brasileira Agroecologia, v.2, n.1, p. 511-514, feb. 2007.

NARDINI, R. C. Socio-environmental diagnosis of the Água Fria stream basin, municipality of Bofete-SP. 2013. 135 f. Thesis (Doctorate in Agronomy) - Universidade Estadual Paulista, Faculdade de Ciências Agronômicas, Botucatu-SP, 2013.

NASUTI, S.; EIRÓ, F.; LINDOSO, D. The challenges of agriculture in the Brazilian semi-arid region. Sustainability in debate. v.4, n.2, p.276-298, jul./dez. 2013.

NETO, J. P. C. et al. Environmental degradation and socioeconomic conditions in the municipality of Vitória do Mearim-Maranhão. Revista Econômica do Nordeste, v. 39, n.2, p. 302-327, Apr./Jun. 2008.

NEY, M. G.; PEREIRA. F. da S.; ZAMPIROLLI, P. D. A. A importância da educação para a reforma agrária e a melhoria do nível de renda dos empregados na agricultura brasileira. In: Congress of the Brazilian Society of Economics, Administration and Rural Sociology, 47, 2009. Porto Alegre-RS. Proceedings... Porto Alegre-RS: SOBER, 2009.

OLIVEIRA, E. da S. The importance of final disposal of empty pesticide containers. UNIABEU, v.5, n.11, p. 123-135, Sept./Dec. 2012.

OLIVEIRA, R. R. de; BARROS, J. D. de S.; SILVA, M. de F. P. da. Desertification and environmental degradation: farmers' perceptions in the municipality of Cachoeira dos Índios/PB. Polêm!ca, v. 11, n. 2, p. 244-251, apr./jun. 2012.

OLIVEIRA, D. B. S. de. The use of social water technologies in rural areas of the semi-arid region of Paraíba: between combating drought and living with the semi-arid region. 2013. 168 f. Dissertation (Master's in Geography) - Federal University of Paraíba, João Pessoa-PB, 2013.

PATRÍCIO, P. C.; GOMES, J.C.C Sustainable rural development, planning and participation. Revista NERA, n.21, p. 100-113, Jul./Dec. 2012.

PENTEADO, S. R. Organic Agriculture. In:. Introduction to Organic Agriculture. Viçosa: Aprenda Fácil, 2003. p. 15-32.

PEREIRA, L. A. et al. Agriculture and its relationship with the environment. In: BRITO, L. T. de L.; MELO, R. F. de (Orgs). Environmental impacts caused by agriculture in the Brazilian semi-arid region. Petrolina: Embrapa Semiárido, 2010. p. 13-29.

PEREIRA, R. A.; BARBOSA, M. de F. N. Socio-economic and environmental diagnosis of a
watershed in the semiarid region of Paraiba. Environmental Engineering, v. 6, n. 1, p. 137153, 2009.

PISANI, R. J. et al. Socio-economic and environmental diagnosis as a planning tool for family farming. Case study: Rio das Pedras sub-basin, Itatinga-SP. Caminhos de Geografia, v.12, n.40, p. 70-79, dec. 2011.

PRODONOV, C. C. Metodologia do trabalho científico: métodos e técnicas da pesquisa e do trabalho acadêmico. Novo Hamburgo: Feevale, 2013.

QUEIROGA, R. A. et al. Socioeconomic and Environmental Diagnosis of the Veneza Settlement, Municipality of Aparecida, Paraíba/PB. Agropecuária Científica no Semiárido, v.10, n.4, p. 05-14, Oct./Dec. 2014

ROCHA, J. S. M. da. Environmental project manual. Santa Maria: UFSM, 1997. 423p.

ROSA, P. V.; FREIRE, J. M. Agroecology: scientific knowledge and/or popular knowledge? Breves Contribuciones I.E.G, n.22, p. 166-193. 2010/2011.

SÁ, I. B.; ANGELOTTI, F. Environmental degradation and desertification in the Brazilian semi-arid region. In: ANGELOTTI, F.; SÁ, I. B.; MENEZES, E. A.; PELLEGRINO, G. Q. Mudanças climáticas e desertificação no Semi- Árido brasileiro. Petrolina - PE: Embrapa Semiárido, 2009. p.53-76.

SANTANA, E. P. R. S; OLIVEIRA, A. R; OLIVEIRA, F. J. M. Diagnóstico socioeconômico da comunidade de Pindoba, Município de Areia - PB. Revista verde, v. 3, n.4, p. 46-62, Oct/Dec, 2008.

SANTOS, J. M. dos. Coexistence strategies for conserving natural resources and mitigating the effects of desertification in the semi-arid region. In: LIMA, R. da C. C.; CAVALCANTE, A. de M. B.; MARIN, A. M. P. (Orgs). Desertification and climate change in the Brazilian semi-arid region. Campina Grande: INSA-PB, 2011. p. 163-184.

SANTOS, A. M. dos; MARÇAL, N. A.; PINTO, E. do N. F. Organic production guaranteeing the health of rural workers. Revista Brasileira de Gestão Ambiental, v.8, n.1, p. 01-05, jan./mar. 2014.

SANTOS, C. F. dos. Diagnosis of family farming in the municipality of Janduís/RN: social, economic and environmental perspective. 2013. 102 f. Dissertation (Postgraduate Programme in Environment, Technology and Society) -

Federal Rural University of the Semi-Arid, Pro-Rectory of Research and Postgraduate Studies, Mossoró-RN, 2013.

SANTOS, J. O. dos et al. The Evolution of Organic Agriculture. Revista Brasileira de Gestão Ambiental, v.2, n.1, p. 35-41, Jan./Dec. 2012.

SILVA, R. M. A. da. Between two paradigms: combating drought and living in the semi-arid region. Sociedade e Estado, v.18, n.1/2, p. 361-385, jan./dez. 2003

. Between combating drought and living with the semi-arid region: paradigmatic transitions and the sustainability of development. 2006. 298 f. Thesis (PhD in Sustainable Development) - Centre for Sustainable Development, University of Brasília, Brasília-DF, 2006.

SILVA, D. D. E. da; RIOS, F. R. de A. Environmental degradation: an analysis of agriculture in the semi-arid Northeast. Revista Brasileira de Gestão Ambiental, v.7, n.2, p. 01-06, Apr./Jun. 2013.

SILVA, P. C. G. da et al. Characterisation of the Brazilian semi-arid region: natural and human factors. In: SÁ, I. B.; SILVA, P. C. G. da (Orgs). The Brazilian semi-arid region: research, development and innovation. Petrolina: Embrapa Semiárido, 2010. p. 17-48.

SILVA, J. S. Agroecology: a strategic basis for food security. Revista Verde de Agroecologia e Desenvolvimento Sustentável, v.5, n.1, p. 01-06, jan./mar. 2010.

SILVA, B. C. D. da; COSTA, A. E. D. V. Socio-productive diagnosis of family farmers co-operating with the family farmer co-operative in the territory of the recôncavo of Bahia. Magistra, v. 24, n. 2, p. 151-159, apr./jun. 2012.

SILVA, F. M. da. et al. The risks of indiscriminate use of pesticides: a bibliographical overview. INTESA, v. 9, n. 1, p. 77-84, jan./jun. 2015.

SILVA, D. D. C.; MATTOS, A. Socioeconomic and environmental diagnosis in a watershed located in a desertification centre. Caminhos de Geografia, v. 14, n.45, p. 45-53, 2013.

SOUSA,V. G. de. Socio-economic and environmental diagnosis and prognosis of the springs of the

Riacho das Piabas (PB). 2010. 108f. Dissertation (Master's Degree in Natural Resources) -
Federal University of Campina Grande. Campina Grande - PB, 2010.

SOUZA, U. V. de. Social technologies as tools for educommunication and the production of discursive and imagetic content about the Brazilian semi-arid region: an experience report from social organisations in conjunction with ASA. Revista ComSertões, v.1, n.2, not paginated. 2014.

SOUZA, J. L. de. Agroecology and organic farming: principles, methods and practices. 2 ed. Vitória: Incaper, 2015, 34p.

SOUZA, G. A. V. da S. et al. Socioeconomic and environmental diagnosis of family farmers in Sítio Jardim, Areia-PB. Cadernos de Agroecologia, v.6, n.2, p. 1-4, dec. 2011.

SOARES, I. F.; MELO, A. C. de; CHAVES, A. D. C. G. Family farming: an alternative for sustainable development in the municipality of Condado-PB. INFOTECNARIDO, v.3, n.1, p. 56-63, jan./dez. 2009.

TORRES, J. L. R.; VIEIRA, D. M. da S. Socioeconomic, environmental and morphological analysis of the micro-basin of the Pintos stream, a tributary of the Uberaba River. Enciclopédia Biosfera, v.9, n.16, p. 243-258, 2013.

VÁSQUES, S. F.; BARROS, J. D. de S.; SILVA, M. de F. P. da. Alternatives to conventional agriculture. Revista Verde, v.3, n.3, p. 06-12, jul./set. 2008.

ZAMBERLAM, J.; FRONCHETI, A. Agricultura ecológica: preservação do pequeno agricultor e do meio ambiente. 3 ed. Petropólis: Vozes, 2007. 213 p.

. Agroecology: the way to preserve farmers and the environment. Petrópolis: Vozes, 2012. 196p.

APPENDICES

Appendix A. Socio-economic and environmental indicators

SOCIAL FACTOR

a) Demographic Variable

Chart 01 - Socio-economic diagnosis - codes and stratification criteria, social factor, demographic variable.

Code 1.1: Age of head of household

ALTERNATIVES	VALUES POINTERED
21-25	1
25-30	2
31-35	3
36-40	4
41-45	5

46-50	6
51-55	7
56-60	8
61-65	9
66 or < 20	10

Code 1.2: Level of education of the head of household

ALTERNATIVES	VALUES POINTERED
Graduation/Specialisation/Master's Degree/ Doctorate/Lecturer	1
Complete high school or technical course	2
High school incomplete	3
Primary school II	4
Primary school 1	8
Illiterate	10

Code 1.3: Average age of the family unit

ALTERNATIVES	WEIGHTED VALUES
21-25	1
25-30	2
31-35	3
36-40	4
41-45	5
46-50	6
51-55	7
56-60	8
61-65	9
£ 66 or < 20	10

Code 1.4: Total number of people in the family nucleus (head and wife + children)

ALTERNATIVES	WEIGHTED VALUES
3-4 people	1
5-6 people	3
7-8 people	4
1 -2 people	7
Up to 9 people	10

b) Housing variable

Table 02 - Socio-economic diagnosis - codes and stratification criteria, social factor, housing variable.

Code 2.1: Type of dwelling

ALTERNATIVES	WEIGHTED VALUES
House of any kind great	1
Good masonry house	2
Bad masonry house	3
Brick and rammed earth house	4
Good rammed earth house (good pau a pique)	6
Poor rammed earth house (pau a pique ruim)	8
Tin/cardboard/straw house	10

Code 2.2: Number of rooms r home

ALTERNATIVES	WEIGHTED VALUES
>10 pieces	1
8-9 pieces	2
6-7 pieces	4
4-5 pieces	7
2-3 pieces	8
1 piece	10

Code 2.3: Average number of people per room

ALTERNATIVES	WEIGHTED VALUES
1 person	1
2 people	2
3 people	4
4 people	6
5 people	8
More than 5 people	10

Code 2.4: Type of cooker

ALTERNATIVES	WEIGHTED VALUES
(Electric and/or biogas and/or microwaves) + Gas	1
Gas	2
Gas and wood/coal	4
Coal/wood	10

Code 2.5: Water for human consumption

ALTERNATIVES	WEIGHTED VALUES
Potable = filtered + (boiled or chlorinated or SODIS)	1
Non-potable	10

Code 2.6: Origin of water consumed for human consumption

ALTERNATIVES	WEIGHTED VALUES
Public network	1

Well/fresh water	2
Spout/cistern	3
Cistern	4
Weir/river/stream/barrier	6
Water tanker	10

Code 2.7: Sewage

ALTERNATIVES	WEIGHTED VALUES
Sewerage system	1
Black well or septic tank	3
Free disposal	10

Code 2.8: Waste disposal

ALTERNATIVE	WEIGHTED VALUES
Selective collection	1
Public collection	2
Bury or burn	5
Free	10

Code 2.9: Disposal of pesticide packaging

ALTERNATIVES	WEIGHTED VALUES
Triple-washing and collection of packaging for the selling companies themselves	1
Any other use or destination of the packaging	10

Code 2.10: Domestic appliances and electronics

ALTERNATIVES	WEIGHTED VALUES
Over 10 items	1
Between 7 and 9 items	2
Between 5 and 6 items	3
Between 3 and 4 items	6
Between 1 and 2 items	8
None	10

c) Variable Availability of Ali ments

Table 03 - Socioeconomic diagnosis stratification, social factor, variable d economic - food availability codes and criteria.

Code 3.1 to 3.15: Availability of and food

CODE 0	ALL ITEMS	Weighted value 0 (award)	DAYS/WEEK	ALTERNATIVES AS	V. P.
3.1	Consumption of milk and dairy products		7	Very high	1

	(cheese , yoghurt, curd, dulce de leche)				
3.2	Meat consumption (cattle, pork or game)		6	High	2
3.3	Fruit consumption		5	Medium High	3
3.4	Consumption of		4	Medium	4
	vegetables				
3.5	Sweet potato/maple/yam consumption		3	Medium Low	5
3.6	Egg consumption		2	Bass	6
3.7	Pasta consumption		1	Very low	7
3.8	Consumption of rice and/or beans		none	-	1 0
3.9	Fish consumption				
3.10	Poultry/hunted game consumption				
3.11	Coffee/tea consumption				
3.12	Consumption of couscous and other maize derivatives: cake/angu/xerém/mungu nzá				
3.13	Consumption of bread/biscuits/wheatcake				
3.14	Consumption of rapadura/sweet				
3.15	Consumption of manioc/tapioca flour e derivatives				

d) Variable Participation in Organisation (Association)

Table 04 - Socio-economic diagnosis - codes and stratification criteria, social factor, participation in organisation variable.

Code 4.1: Membership of an organisation (association)

ALTERNATIVES		Yes or no	CONSIDERATIO N	WEIGHTED VALUES
If you use	use of		All 08 items (yes)	1

collective machinery/equipment			
If you take part in any community/collective projects		7 items (yes)	2
If you are or have been a board member		6 items (yes)	3
If he attends meetings		5 items (yes)	4
If he knows the statute		4 items (yes)	5
If you are part of an association		3 items (yes)	6
If you are part of a co-operative		2 items (yes)	7
If you are a union member		Just one item	8
You've been part of		-	9
Not part of it		-	10

e) Rural health variable

Table 05 - Socio-economic diagnosis - codes and stratification criteria, social factor, rural health variable.

Code 5.1: Pest infestation (attacks by nematodes, termites, ants, grasshoppers, caterpillars, ectoparasites, mealybugs, rats, flies, fleas, mosquitoes, lice, cockroaches and animal vermin)
Other:

ALTERNATIVES	WEIGHTED VALUES
Nil	1
Low	3
Average	5
High	7
Impediment	10
NULL - No infestation LOW - Small infestation MEDIUM - Infestation of medium severity HIGH - Heavy and extensive infestation IMPEDITIVE - Infestation so great as to make it impossible to exploit the land	

Code 5.2: Pest control

ALTERNATIVES	WEIGHTED VALUES
IPM (Integrated Pest Management)	1
Biological (enemy plants and natural enemies)	2
Systematic (periodic)	3

Eventual	5
Never	10

Code 5.3: Human health

ALTERNATIVES	WEIGHTED VALUES
Great	1
Regular	3
Low	5
Bad	7
Inhospitable	10
Note: Environmental conditions affect the well-being and health of plants, livestock and humans, especially with regard to temperature, relative humidity and the occurrence of endemic diseases and pests, such as impaludence, anaemia, schistosomiasis, chagas disease, lice infestation, dengue fever, environmental dirt, among others. GREAT - Easy human labour, no heat, good relative humidity, no endemic diseases REGULAR - Mild temperature and relative humidity, presence of endemic diseases LOW - High temperature and relative humidity, endemic infestations BAD - Excessively hot weather and high relative humidity or excessively dry, dirty environment with endemic infestation INOSPITE - Excessively hot climate (high temperature) and high humidity, filthy environmental aspect, with endemic infestation	

f) Variable Application of Laws (when people work on the property outside the family throughout the year)

Chart 06 - Socio-economic diagnosis - codes and stratification criteria, social factor, variable application of laws.

Code 6.1: Child labour

ALTERNATIVES	WEIGHTED VALUES
It doesn't exist	1
There is	10

Code 6.2: Working regime (daily rural working time 8 hours)

ALTERNATIVES	WEIGHTED VALUES
It doesn't exist	10
There is	1

ECONOMIC FACTOR

a) Variable Production

Table 07 - Socio-economic diagnosis - codes and stratification criteria, economic factor, production variable.

Code 7.1: Variable average agricultural productivity

ALTERNATIVES	WEIGHTED VALUES
Above average	1
On average	2
Below average	5
Does not produce	10

Main types of crops to consider: maize, potatoes, sorghum, sunflower, cotton, manioc, beans, vegetables, sugar cane, tomatoes, onions, vegetables in general, fruit in general, etc.

Code 7.2 and 7.3: Afforestation and planted pastures

CODE	ALTERNATIVES		WEIGHTED VALUES
7.2	Afforestation (inclu de native forest)/arborisation	£ 20% of the area	1
		10-19% of the area	5
		1 -9% of the area	8
		Below 1% / None	10
7.3	Planted pastures planted (grass, palm, pasture grass)	Preserved pasture + silage/pennage	1
		Conserved pasture without strategic food reserves	3
		Degraded pasture invaded by weeds or pioneers	5
		Purchase of extra bulk	8
		property	
		None	10

b) Variable Working Animals

Table 08 - Socio-economic diagnosis - codes and stratification criteria, economic factor, working animal activity variable.

Code 8.1: Labour animals variable

ALTERNATIVES	WEIGHTED VALUES
Ox + horse + donkey/burro (03 animals used for rural work - transporting produce, ploughing...)	1
Only two of them	3
Only one of them	5
None of them	10

c) Variable Production Animals

Table 09 - Socio-economic diagnosis - codes and stratification criteria, economic factor, livestock production activity variable.

Code 9.1: Livestock variable

ALTERNATIVES	WEIGHTED VALUES
It has > 4 types of production animals	1
It has 3 types of livestock	3
It has 2 types of production animals	5
It has 1 type of production animal	7
No type of animal	10

d) Variable Commercialisation, Credit and Income

Table 10 - Socio-economic diagnosis - codes and stratification criteria, economic factor, commercialisation variable, credit and income.

Code 10.1: To whom you sell agricultural produce

ALTERNATIVES	WEIGHTED VALUES
Co-operatives	1
Ceasa	3
Agribusiness	4
Grocery (retail)	5
Consumers	7
Intermediate	8
Not for sale	10

Code 10.2: To whom you sell livestock production

ALTERNATIVES	WEIGHTED VALUES
Co-operatives	1
Refrigerators	3
Animal fair	4
Marchante (retail)	5
Intermediate	7
Consumers	8
Not for sale	10

Code 10.3: To whom you sell forestry production (umbu, charcoal, chestnuts, firewood)

ALTERNATIVES	WEIGHTED VALUES
Consumers	1
Co-operatives	3
Ceasa	4
Agribusiness	5
Grocery (retail)	7

Intermediate	8
Not for sale	10

Code 10.4: Main source of agricultural credit

ALTERNATIVES	WEIGHTED VALUES		
Own resource	1		
Co-operatives	2		
Revolving fund	3	0 producer receives the good and replenishes it in a timely manner	Fund set up with resources
		established to serve another producer	of a group
Official Bank	4		
Agro-industry/Fry	6		
Private banks	8		
Loan shark (private individuals)	9		
No access to credit	10		

Code 10.5: Approximate gross income of the property (monthly)

ALTERNATIVES	WEIGHTED VALUES
> 5 minimum wages	1
4-5 minimum wages	2
3-4 minimum wages	3
2-3 minimum wages	4
1-2 minimum wages	7
14-1 minimum wage	9
Up to ^ minimum wage	10

Code 10.6: Other income

ALTERNATIVES	WEIGHTED VALUES
Has (quote opposite)	1
None	10

Code 10.7: Total income

ALTERNATIVES	WEIGHTED VALUES
> 5 minimum wages	1
>4-5 minimum wage	2
>3-4 minimum wages	3
>2-3 minimum wages	4
>1-2 minimum wages	7
>14-1 minimum wage	9
Up to ^ minimum wage	10

a) Technological Variable

Table 11 - Socio-economic diagnosis - codes and stratification criteria, technological factor, technological variable.

Code 11.1: Property area (in ha)

ALTERNATIVES	WEIGHTED VALUES
More than 200 ha and with utilisation above 50%	1
From 101 to 200 ha and with utilisation above 50%	2
From 21 to 100 ha and with utilisation above 50%	4
Less than 20 ha and with utilisation above 50%	6
More than 20 ha and up to 50% utilisation	8
Less than 20 ha and up to 50% utilisation	10

Code 11.2: Type of ownership

ALTERNATIVES	WEIGHTED VALUES
Owner	1
Land reform tenant	3
Lessee	5
Steward	7
Illegal occupier/landlord	10

Code 11.3: Use of pesticides (fu ngicides, insecticides, herbicides)

ALTERNATIVES	WEIGHTED VALUES
IPM (Integrated Pest Management)	1
Biological control	2
With technical guidance	3
Empirically orientated	8
No combat and/or control	10

Code 11.4: Fertilising and/or liming

ALTERNATIVES	WEIGHTED VALUES
Not used	10
Eventual Chemistry, without technical guidance	8
Chemistry, according to guidelines	3
technique	
Chemistry and organics technical guidance	2
Organic / green manure and crop	1

rotation among other conservation practices	

Code 11.5: Type of tools/implements you have to deal with on the property

ALTERNATIVES	WEIGHTED VALUES
Both	1
Mechanics	5
Manual	10

Code 11.6: Logistics on the farm (type of transport for transporting production)

ALTERNATIVES	WEIGHTED VALUES
Own vehicle	1
Alternative transport	2
Buses	3
Motorbike	4
Animal-drawn cart	6
Horse, donkey, ass...	7
Bicycle	8
Wheelbarrow	9
None	10

Code 11.7: Type of soil preparation for planting, in terms of bed direction

ALTERNATIVES	WEIGHTED VALUES
Planting in contour (terracing)	1
Downhill (in favour of the slope)	10

Code 11.8: Regarding the animal feed reserve

ALTERNATIVES	Yes or No	WEIGHTED VALUES
Silage and/or haymaking practices in the rainy season		4 items (yes) = 1
It has protein crops for animals		3 items (yes) = 3
(sorghum, leucena...)		
It has energy crops (maize, sorghum, etc.)		2 items (yes) = 5
It has voluminous crops such as (palm, weeds...)		1 item (yes) = 7
Does not have any of the above alternatives for animal feed		10

Code 11.9: Soil conservation practices

ALTERNATIVES	Yes or No	WEIGHTED VALUES
Terracing		9 items (yes) = 1
Planting on a contour		8 items (yes) = 2
Intercropping		7 items (yes) = 3
Crop rotation		6 items (yes) = 4
Minimum tillage (no-till and/or light harrow or cultivator for soil preparation or mowing)		5 items (yes) = 5

Mulching (straw, manure and cultural remains on the ground)		4 items (yes) = 6
Leirons on contour lines		3 items (yes) = 7
Living fences		2 items (yes) = 8
Strip planting or agroforestry system		1 item (yes) = 9
No soil conservation practices		No technique = 10

Code 11. 10: Regarding the use of irrigation

ALTERNATIVES	WEIGHTED VALUES
Regular	1
Occasional (supplementary)	5
Not used	10

Code 11.11: Technical assistance and assimilation of guidelines

ALTERNATIVES	WEIGHTED VALUES
Regular and follow the guidelines	1
Occasional and follows the guidelines	3
Regular and not always following the guidelines	5
Occasional and not always followed	7
the guidelines	
No technical assistance	10

Code 11.12: Agricultural land use practices with a high conservation risk

ALTERNATIVES	Yes or No	WEIGHTED VALUES
Annual planting on shallow soils		9 items (yes) = 10
Monoculture		8 items (yes) = 9
Erosion		7 items (yes) = 8
Burning		6 items (yes) = 7
Annual planting in areas with a slope of 15% or more		5 items (yes) = 6
Downhill planting		4 items (yes) = 5
Deforestation above 80% of the total area		3 items (yes) = 4
Absence of riparian forest		2 items (yes) = 3
Degraded soils (no productivity and absent agricultural layer)		1 item (yes) = 2
No practices that do not comply with soil conservation and risk soil degradation		1

Code 11.13: Existence of water collection techniques on the property

ALTERNATIVES	Yes or No	WEIGHTED VALUES
Weirs/dams		9 items (yes) = 1
Cacimba (unlined hole in the riverbed)		8 items (yes) = 2
Amazon or cacimbão wells (lined walls)		= items (yes) = 3
Underground dam		6 items (yes) = 4

Tube well		5 items (yes) = 5
Cisterns for collecting water from roofs or pavements		4 items (yes) = 6
Stone tank		3 items (yes) = 7
Barraginhas		2 items (yes) = 8
Zero base or successive surface bars		1 item (yes) = 9
No form of water collection		No technique
		= 10

Code 11.14: Form of livestock farming

ALTERNATIVES	WEIGHTED VALUES
Intensive (animals in confined paddocks with 80% feed at the trough)	1
Semi-extensive (afternoon pick-up for supplementary feeding)	5
Extensive (animals not put to sleep)	10

b) Variable Machinery and Verticalisation of Production (Rural Industrialisation)

Table 12 - Socio-economic diagnosis - codes and stratification criteria, technological factor, machinery variable and rural industrialisation.

Code 12.1: Owns agricultural machinery and implements (individual or collective use through associations)

ALTERNATIVES	Yes or No	WEIGHTED VALUES
Animal or mechanical traction cultivator		9 items (yes) = 1
Animal-drawn cart or motor-drawn wagon		8 items (yes) = 2
Diesel or electric forage harvester		7 items (yes) = 3
Diesel, electric or manual pump		6 items (yes) = 4
Backpack or mechanised sprayer		5 items (yes) = 5
Manual, animal-drawn or mechanised planter		4 items (yes) = 6
Mechanical milking machine		3 items (yes) =7
Cereal mixer, sherry mill...		2 items (yes) = 8
Any other machine that facilitates work in the field (cite...)		1 item (yes) = 9
No aid machines in the field		No machine = 10

Code 12.2: Adds value by processing wood, fruit, milk, meat, honey, skins, fish and others

ALTERNATIVES	WEIGHTED VALUES
Processing with legal quality standards	1

(technicalised)	
Rough/manual processing	5
There is no processing of the products generated on the property	10

Code 12.3: Some kind of handicraft

ALTERNATIVES	WEIGHTED VALUES
Yes, for sale regularly	1
Yes, for own consumption	5
No	10

GENERAL DIAGNOSES

Table 13 - General diagnoses - priority factor, general variables to be answered by the farmer interviewed (moment of diagnosis identified by the micro-watershed actor himself)

PERSONAL IDENTIFICATION OF THE PRODUCER	Order of priorities (write from 1ª to 4ª priority according to the producer)
Priority problems (tick the first four)	
ALTERNATIVES	
1-Possession of land (property)	
2-Little land	
3-Low production/productivity	
4-Water shortage	
5-Lack of electricity	
6-Lack of sewage	
7-Lack of medical and dental care	
8-Lack of good housing	
9-Lack of credit	
10-Lack of market	
11-Low incomes (poorly valued product or salary)	
12-Roads (bad - lacking)	
13-Technical assistance	
14-Schools	
15-inputs (raw materials, labour force, energy consumption, etc.)	
Other - please quote	

Q.14	ENVIRONMENTAL DIAGNOSIS	Date:	
Code	Polluting Elements (without technical-scientific guidance)	Value found	Observations
14.1	Stocking up on pesticides		
14.2	Pesticide packaging depots		
14.3	Places to wash pesticide application		

	implements		
14.4	Application of pesticides		
14.5	Waste bins (urban, rural waste) - Monturo		
14.6	Logging (firewood, charcoal, stakes, poles, etc.)		
14.7	Wild animal breeding		
14.8	Hunting for sale		
14.9	Pigsties		
14.10	A vi a ri s / sta bu 1 s (coach / car 1)		
14.11	Slaughterhouses (slaughtering animals for sale)		
14.12	Marked erosion (on the ground or in the street/road)		
14.13	Open sewers		
14.14	Burning		
14.15	Pumps for pumping water out of rivers/weirs		
14.16	Whey		
14.17	Use of insecticides with your hands - use of toxin gas (tablets) in bean bags		
14.18	Exploitation of native species		
14.19	Water use for irrigation		

Conventions:

Alternatives	Weighted Value
It doesn't exist	1
There is technical-scientific guidance	2
Exists without technical-scientific guidance	3

Appendix B. Informed Consent Form

FEDERAL UNIVERSITY OF CAMPINA GRANDE TEACHER TRAINING CENTRE ACADEMIC UNIT OF EXACT AND NATURAL SCIENCES DEGREE COURSE IN BIOLOGICAL SCIENCES INFORMED CONSENT FORM

You are being invited to participate as a volunteer in the study ADOPTION OF SUSTAINABLE AGRICULTURAL PRACTICES: CHALLENGES AND POTENTIALITIES IN RURAL COMMUNITIES SUBSIDISED IN THE SERTÃO PARAIBANO BY THE SÃO FRANCISCO RIVER TRANSPOSITION PROJECT, coordinated by Professor Dr. José Deomar Souza Barros and linked to the UNIVERSITY OF THE ACADEMIC ACADEMIC OF EXACT AND NATURAL SCIENCES OF THE TEACHER TRAINING CENTRE OF THE UNIVERSITY OF THE FEDERAL FEDERATION. José Deomar de Souza Barros and linked to the ACADEMIC UNIT OF EXACT AND NATURE SCIENCES OF THE TEACHER TRAINING CENTRE OF THE FEDERAL UNIVERSITY OF CAMPINA GRANDE.

Your participation is voluntary and you can withdraw your consent at any time without any harm or penalty. The aim of this study is to evaluate the agricultural practices adopted in rural communities settled by the São Francisco River transposition project, as well as their social, economic and technological aspects, and it is necessary because it will provide data that could contribute to the development of public policies for the study region. Bearing in mind that until now there has been no research with this focus in the region covered by the study. With the growing need to discuss socio-economic, technological and agro-ecological problems, studies related to sustainability in general are relevant for articulating with society in the current context, taking into account the real socio-economic conditions of individuals and their interaction with the environment.

If you decide to accept the invitation, you will be subjected to the following procedure(s): you will answer the questions contained in the research questionnaire, you will be subjected to observation while carrying out your farming activities. The risks involved with your participation are: you may feel uncomfortable sharing personal or confidential information, or on some topics you may feel uncomfortable talking about, or you may feel embarrassed when carrying out your farming activities. As a mitigating measure, the research subject does not have to answer any questions if they feel they are too personal or feel uncomfortable speaking, and they can also request that at certain times they feel it is appropriate to request that their activities not be observed. The benefits of the research will be: the research will make it possible to diagnose the socio-economic, technological and agro-ecological profile of the local population, generating data that could foster the implementation of public policies aimed at benefiting the local population, with regard to the adoption of sustainable agricultural practices.

All information obtained will be kept confidential and your name will not be identified at any time. The data will be stored in a safe place and the results will be publicised in a way that does not allow any volunteer to be identified.

If you incur any expenses as a result of your participation in the research, you will be reimbursed if you request it. At any time, if you suffer any proven damage as a result of this research, you will be compensated.

You will keep a signed and initialled copy of this form and any questions regarding this research can be addressed to José Deomar de Souza Barros, whose contact details are specified below.

Contact details for the person responsible for the research

Name: José Deomar de Souza Barros
Institution: Federal University of Campina Grande
Address: Rua Sérgio Moreira de Figuiredo S/N - Casas populares. Cajazeiras - PB.
Telephone: (83) 3532 - 2111
Email: deomarbarros@gmail.com

I declare that I am aware of the objectives and importance of this research, as well as the way in which it will be conducted, including the risks and benefits related to my participation, and I agree to voluntarily participate in this study.

Cajazeiras - PB, 23rd February 2015

Signature or fingerprint of volunteer

José Deomar de Souza Barros

Printed by Books on Demand GmbH, Norderstedt / Germany